ADVANCES IN IMAGING AND
ELECTRON PHYSICS
PARTIAL CUMULATIVE INDEX

VOLUME 100

EDITOR-IN-CHIEF

PETER W. HAWKES
*CEMES/Laboratoire d'Optique Electronique
du Centre National de la Recherche Scientifique
Toulouse, France*

ASSOCIATE EDITORS

BENJAMIN KAZAN
*Xerox Corporation
Palo Alto Research Center
Palo Alto, California*

TOM MULVEY
*Department of Electronic Engineering and Applied Physics
Aston University
Birmingham, United Kingdom*

Advances in
Imaging and Electron Physics
Partial Cumulative Index

EDITED BY
PETER W. HAWKES

*CEMES/Laboratoire d'Optique Electronique
du Centre National de la Recherche Scientifique
Toulouse, France*

VOLUME 100

ACADEMIC PRESS
San Diego London Boston New York
Sydney Tokyo Toronto

This book is printed on acid-free paper. ∞

Copyright © 1998 by ACADEMIC PRESS

All Rights Reserved.
No part of this publication may be reproduced or transmitted in any form or by any means, electronic or mechanical, including photocopy, recording, or any information storage and retrieval system, without permission in writing from the publisher.
The appearance of the code at the bottom of the first page of a chapter in this book indicates the Publisher's consent that copies of the chapter may be made for personal or internal use, or for the personal or internal use of specific clients. This consent is given on the condition, however, that the copier pay the stated per copy fee through the Copyright Clearance Center, Inc.
(222 Rosewood Drive, Danvers, Massachusetts 01923), for copying beyond that permitted by Sections 107 or 108 of the U.S. Copyright Law. This consent does not extend to other kinds of copying, such as copying for general distribution, for advertising or promotional purposes, for creating new collective works, or for resale. Copy fees for pre-1997 chapters are as shown on the chapter title pages; if no fee code appears on the chapter title page, the copy fee is the same as for current chapters.
1076-5670/98 $25.00

ACADEMIC PRESS
525 B Street, Suite 1900, San Diego, CA 92101-4495, USA
1300 Boylston Street, Chestnut Hill, MA 02167, USA
http://www.apnet.com

Academic Press Limited
24-28 Oval Road, London NW1 7DX, UK
http://www.hbuk.co.uk/ap/

International Standard Book Number: 0-12-014742-4

PRINTED IN THE UNITED STATES OF AMERICA
97 98 99 00 01 IC 9 8 7 6 5 4 3 2 1

VOLUMES INCLUDED IN THIS INDEX

Advances in Optical and Electron Microscopy

VOLUME 1	1966
VOLUME 2	1968
VOLUME 3	1969
VOLUME 4	1971
VOLUME 5	1973
VOLUME 6	1975
VOLUME 7	1978
VOLUME 8	1982
VOLUME 9	1984
VOLUME 10	1987
VOLUME 11	1989
VOLUME 12	1991
VOLUME 13	1994
VOLUME 14	1994

Advances in Electronics and Electron Physics

VOLUME 82	1991
VOLUME 83	1992
VOLUME 84	1992
VOLUME 85	1993
VOLUME 86	1993
VOLUME 87	1994
VOLUME 88	1994
VOLUME 89	1994

Advances in Imaging and Electron Physics

VOLUME 90	1995
VOLUME 91	1995
VOLUME 92	1995
VOLUME 93	1995
VOLUME 94	1995
VOLUME 95	1996
VOLUME 96	1996
VOLUME 97	1996
VOLUME 98	1996
VOLUME 99	1997

PREFACE

In order to celebrate the 100th volume of these Advances, a cumulative index of the volumes that have appeared since the last thematic index was published, in volume 81, has been prepared. Originally, I hoped to provide an index of the entire 99 volumes of the serial that began as *Advances in Electronics*, then soon became *Advances in Electronics and Electron Physics* and, since volume 90, has been *Advances in Imaging and Electron Physics*. The satellite series *Advances in Image Pick-up and Display* (edited by Ben Kazan) and *Advances in Optical and Electron Microscopy* (first edited by V. E. Cosslett and R. Barer and subsequently by Tom Mulvey) were likewise to have been included. It rapidly became apparent that such an index would be far too voluminous and the present index is therefore confined to volumes 82–99 of AEEP/AIEP but does include the whole of *Advances in Optical and Electron Microscopy*. We plan to produce indexes of themes and contributors, similar to those to be found in volume 81 but covering all three series and the occasional supplementary volumes, in the coming year.

I am very pleased to report that numerous articles are promised for the second hundred volumes of the series. A list of these follows.

Peter W. Hawkes

FORTHCOMING CONTRIBUTIONS

Nanofabrication	H. Ahmed and W. Chen (vol. 102)
Finite-element methods for eddy-current problems	R. Albanese and G. Rubinacci (vol. 102)
Mathematical models for natural images	L. Alvarez Leon and J. M. Morel
Use of the hypermatrix	D. Antzoulatos
Image processing with signal-dependent noise	H. H. Arsenault
The Wigner distribution	M. J. Bastiaans
Hexagon-based image processing	S. B. M. Bell
Modern map methods for particle optics	M. Berz and colleagues
Magneto-transport as a probe of electron dynamics in semiconductor quantum dots	J. Bird
ODE methods	J. C. Butcher
Electron microscopy in mineralogy and geology	P. E. Champness (vol. 101)
Microwave tubes in space	J. A. Dayton

Fuzzy morphology	E. R. Dougherty and D. Sinha
The study of dynamic phenomena in solids using field emission	M. Drechsler
Gabor filters and texture analysis	J. M. H. Du Buf
Miniaturization in electron optics	A. Feinerman (vol. 102)
Liquid metal ion sources	R. G. Forbes
X-ray optics	E. Förster and F. N. Chukhovsky
The critical-voltage effect	A. Fox
Stack filtering	M. Gabbouj
Median filters	N. C. Gallagher and E. Coyle
The development of electron microscopy in Spain	M. I. Herrera and L. Brú
Space-time representation of ultra-wideband signals	E. Heyman and T. Melamed (vol. 103)
Structural analysis of quasicrystals	K. Hiraga (vol. 101)
Number-theoretic transforms and image processing	A. G. J. Holt and Dr. S. Boussakta
Formal polynomials for image processing	A. Imiya (vol. 101)
Contrast transfer and crystal images	K. Ishizuka
Optical interconnects	M. A. Karim and K. M. Iftekharuddin (vol. 102)
External optical feedback effects in semiconductor lasers	M. A. Karim and M. F. Alam
Numerical methods in particle optics	E. Kasper
Surface relief	J. J. Koenderink and A. J. van Doorn
Spin-polarized SEM	K. Koike
Sideband imaging	W. Krakow
Dyadic Green's function microstrip circulator theory	C. M. Krowne (vol. 103)
Vector transformation	W. Li
SEM image processing	N. C. MacDonald
The dual de Broglie wave	M. Molski (vol. 101)
Electronic tools in parapsychology	R. L. Morris
Z-contrast in the STEM and its applications	P. D. Nellist and S. J. Pennycook (vol. 103)
Phase-space treatment of photon beams	G. Nemes
Aspects of mirror electron microscopy	S. Nepijko (vol. 102)

Image processing and the scanning electron microscope	E. Oho
Representation of image operators	B. Olstad
Fractional Fourier transforms	H. M. Ozaktas
HDTV	E. Petajan
Scattering and recoil imaging and spectrometry	J. W. Rabalais
The wave-particle dualism	H. Rauch
Digital analysis of lattice images (DALI)	A. Rosenauer
Electron holography	D. Saldin
X-ray microscopy	G. Schmahl
Accelerator mass spectroscopy	J. P. F. Sellschop
Applications of mathematical morphology	J. Serra
Set-theoretic methods in image processing	M. I. Sezan
Focus-deflection systems and their applications	T. Soma
Electron gun system for color cathode-ray tubes	H. Suzuki
Study of complex fluids by transmission electron microscopy	I. Talmon
New developments in ferroelectrics	J. Toulouse
Organic electroluminescence-materials and devices	T. Tsutsui and Z. Dechun
Electron gun optics	Y. Uchikawa
Very high resolution electron microscopy	D. van Dyck
Morphology on graphs	L. Vincent
Structure, fabrication, and performance of color CRTs	E. Yamazaki
Analytical perturbation methods in charged-particle optics	M. I. Yavor (vol. 103)

ADVANCES IN IMAGING AND ELECTRON PHYSICS
PARTIAL CUMULATIVE INDEX

VOLUME 100

INDEX

Key: Number preceding period indicates volume; number(s) following period indicate pages.

A

AAS *See* Atomic absorption spectroscopy
Abbe, Ernst, 12.244
Abbe's theory, 5.243, 244, 245, 286; 8.3–4, 20, 22; 13.268–272; 91.276
    classical experiments, 8.14–17
    lateral diffraction limit, 14.220
ABCD law, 93.186
Abelian group, 93.3–6
    duality, 93.3, 7, 8
    extended Cooley–Tukey fast Fourier transforms, 93.49–50
    Fourier transform, 93.14–15
    vector space, 93.7–8
Aberration coefficients:
    electrostatic anode lens, 8.243, 246, 247–248, 257–258
    for HR-LVSEMs, 83.229
Aberrations, 5.163, 165, 167, 187, 211, 214, 216, 217, 244, 245, 287, 288, 298
    alignment, 13.195, 196
    canonical *See* Canonical aberration
    characteristics, 14.322–323
    chromatic *See* Chromatic aberration
    combined, 91.30–33, 34; 97.360–361
    correction of, 12.69
    determination by ray tracing, 13.206–207
    electron gun, 8.144–147
    electron off-axis holography, 89.8, 9, 21, 48
    electron probe size, 13.195–199
    in electron trajectories, 13.189–199
    geometrical third-order, 13.191–194
    lenses, 8.246–247, 256; 10.230–232; 12.29, 65; 83.207, 221, 222, 229, 230
    electron lens, 5.216; 83.207, 221–222, 224, 228–232
    scanning transmission electron microscopy, 93.86–87
    in microscope objective, 14.257–270
    astigmatism, 14.260–265
    coma, 14.266–267
    curvature of field, 14.265–266
    distortion, 14.267
    sine condition in, 14.258–259
    spherical, 14.258–259
    mirror-bank energy analyzers, 89.403–410
    equations, 89.397–399
    object width, 89.408–410
    transaxial mirrors, 89.437–441
    monochromatic *See* Monochromatic aberration
    multislice approach, 93.207–215
    rotationally symmetric third-order in electron optical transfer function, 13.280–284
    scanning transmission electron microscopy, 93.86–87

Aberrations, *continued*
  spherical *See* Spherical aberration
  in superconducting electron lens, 5.211, 214, 216, 234
  wave aberrations, 89.8, 9, 21, 48
Aberration theory, eikonal functions, 91.1–35
Abingdon Cross benchmark, 90.383–389, 425
Abrikosov vortex lattice, 4.248; 87.174–175
Absolute center, 90.86
Absorbing potential, 86.186
Absorption:
  band-to-band, 82.248
  edge, 82.246
  free-carrier, 82.255
  free-carrier-induced, 82.248
  from biological macromolecules, 7.282
  impurity-induced, 82.248
  relaxation phenomena, 84.321
Absorption coefficient, 2.182, 184, 188
  for band-to-band transitions, 82.249
  due to free carriers, 82.249
  observed, 82.248
Absorptive atomic scattering factor, 90.217
Abstract algebra, covariance estimation and, 84.262–313
Accelerators, optics, 97.337
Accelerometer, 85.2–4
Access function, 88.69
Access transistor, 86.3
Accuracy of approximation, 94.182
Acetylcholine:
  in motor end plate, 6.204, 206
  as neurotransmitter, 6.206
Acetylcholinesterase:
  acetylthiocholine technique, 6.200, 220
  distribution, 6.203, 206
  function, 6.200
  histochemistry, 6.196, 198
  localization, 6.195, 204
  occurrence, 6.200, 206, 208
Achromat microscope objective, 14.277–278, 305–309

Achromobacter, 5.301, 302
Acid arylsulfatase, 6.181, 183
Acid hydrolase, 6.183, 184, 220
Acid phosphatase, 6.176, 183
  distribution, 6.183
  electron microscopy, 6.180, 181, 182
  Gomori technique, 6.180, 191
  optical microscopy, 6.181, 183
  pH, 6.180
  sources, 6.180
  specificity, 6.182
Acoustic delay line, 11.157
Acoustic impedance, 11.155
Acoustic micrographs, 11.153
Acoustic phase microscope, 11.158
Acoustic reflectivity, 11.163
Acoustic tissue models, 84.321
Acoustic (tissue) parameters, 84.323, 344
Acoustic transducers, 11.154
Acoustic velocity, 11.154
Acoustic vibrations, 11.11
Acousto-optical cells, optical symbolic substitution, 89.81–82
Acousto-optic scanning, 10.70
Acoustospectrography, 84.323, 345
Acridine orange (AO):
  dye binding to DNA, 8.53
  purification, 8.58
Acriflavine (AFL), metachromasy of fluorescence, 8.53
Acrolein, 2.273, 276
ACRONYM (software), 86.92, 154, 156–157
Actin:
  electron micrographs, 7.352
  filament, three-dimensional reconstruction, 7.353
  myosin complex, reconstruction, 7.353
  tropomyosin complex, reconstruction, 7.354
Activation energy, 86.24, 56, 62, 74
  of base current, 82.203
Active data model of computation, 87.294–295

Active matrix addressed display, flat panel technology, 91.248–251
Acyclic graphs, minimax algebra, 90.52–54
Adams–Bashforth formula, 13.202
Adams–Moulton formula, 13.202
Adaptation, in linear prediction, 82.129–130
Adaptive contrast enhancement, 92.34–36
Adaptive extremum sharpening filter, 92.44–48
Adaptive pattern classification, 84.265–266, 270–271
Adaptive postfiltering, 82.170
Adaptive predictive coding, 82.154
Adaptive quantile filter, 92.14–16
Adaptive transform coding, 82.159
Adatoms:
    crystal-aperture scanning transmission electron microscopy, 93.90, 94–100
    electron microscopy, 11.80
Addition:
    linear-time rational calculation, 90.106
    maxpolynomials, 90.88–89
    Minkowski addition, 89.328
    optical symbolic substitution, 89.59–64, 70
Additive property, 91.108
Additive reprojection, 85.139, 146
Additivity, Fisher information, 90.190–191
Adenosine crystals, 5.314
Adenovirus type 5, human, 6.232, 240
    capsomeres, 6.267, 269
    concentrated preparation, 6.265, 267
    disruption, 6.267
    electron micrograph, 6.266
    interference pattern, 6.269
    optical diffraction pattern, 6.266
    recrystallization, 6.267, 268
    structure, 6.240
Adenyl cyclase, 6.190
    histochemical staining, 6.194
    histochemistry, 6.194, 195
    localization, 6.195, 220
Adhesion mounting methods, 8.107

Adiabaticity, 12.122, 125, 130, 133
Adjacency, 85.116
    graph, 84.198–200
    relation, 84.198, 232
Adjugate matrix, minimax algebra, 90.115
Adjustment, 86.225, 234
Adsorbate effects, field electron emissions, in, 8.220
Adsorbate structures, 11.81
Adsorption, 2.345, 365, 401
AEG *See* Allgemeine Elektrizitäts–Gesellschaft
AEI Research Laboratories, 91.269, 279–282
    commercial microscope production, 96.453–457, 463–464
    Corinth, 96.275, 498
    electron diffraction camera production, 96.488–489
    electron microprobe production, 96.464–465
    EM 1, 10.222–223
    EM 6B, 96.474–476
    EM 6G, 96.478
    EM 7, 96.492–493, 531
    EM 801, 96.480–481
    EMMA-4, 96.489–491
    scanning electron microscope research, 96.468
    withdrawal from microscope production, 96.528–529, 531–532
Aerial photography, stripe photomicrography, 7.81, 82
Aerodynamics, 85.80
Aerosols:
    energetic aerosols, 10.206
    holographic analysis, 10.198–201
    study, by in-line holography, 10.141
    wet replication study, 8.77, 81–83, 99
AES *See* Auger electron spectroscopy
Affine group, 93.10–11
    fast Fourier transform algorithm, 93.3
    Cooley–Tukey, 93.46–47
    reduced transform algorithm, 93.30–31, 39–41

Affine group, *continued*
  point group, 93.31–39
  XSUP#sup invariant, 93.41–42
Affinity, 99.41
AFM *See* Atomic force microscope
Africa, Southern:
  electron microscopy:
    andrology, 96.333
    botany, 96.331–332
    ceramics, 96.341
    dentistry, 96.332
    earth science, 96.341–342
    electron microprobes, 96.341–342
    embryology, 96.332–333, 335
    facilities, 96.323–325, 343
    future, 96.343–344
    instrumentation, 96.330–331
    medicine, 96.332–336
    physical sciences, 96.336–341
    plastic deformation, 96.338–341
    replica technique, 96.338
    semiconductors, 96.337
    specimen preparation, 96.331
    thin films, 96.336–337
    virology, 96.334–335
  Electron Microscopy Society of Southern Africa, conferences, 96.323, 325–330
  language barrier, 96.325
Agar filtration sample preparation, 8.114
AGC *See* Automatic gain control
A- and G-divergence, *M*-dimensional generalization, 91.40, 79–82
Agent, reasoning, 86.139, 160–161, 167–168
Aggregate formation, biological particles, 8.114
Agmon–Motzkin–Schoenberg algorithm, 95.178
Aharonov–Bohm effect, 5.252, 253; 89.144–146; 93.174, 176, 178–181
  devices based on, 89.106, 142–178
  electron holography, 99.172, 187–192
  electrostatic, 89.145, 146, 157–178; 99.187–192
  magnetostatic, 89.146, 149–156, 166
  mesoscopic physics, 91.215, 218–219
Aharonov–Bohm interferometer, 89.172–178, 180, 186, 199
Air, glow discharge in, 8.127
Airy disc, 2.48, 49, 55; 10.17, 18
Airy formula, 13.267
Albumin, fixation, 2.289, 2.293 et seq.
ALE *See* Atomic layer epitaxy
Algebra:
  associative, 84.271, 311
  Boolean, fuzzy set theory, 89.258
  Clifford, 84.277
  defined, 84.311
  efficient rational algebra, 90.99–119
  group algebra *See* Group algebra
  image algebra *See* Image algebra
  Jordan algebra *See* Jordan algebra
  Lie, 84.262, 268; 91.2–3
  max algebra *See* Max algebra
  min algebra *See* Min algebra
  minimax algebra *See* Minimax algebra
  spacetime *See* Spacetime algebra
  von Neumann real, 84.272, 284
Algebraic geometry, 86.149
Algebraic product, fuzzy set theory, 89.261–262
Algebraic reconstruction (ART), 95.177, 183
  image reconstruction, 97.160–162
  three-dimensional images, 7.324–327
ALGOL, 5.45
Algorithms, 93.2–3
  adaptive algorithm, 85.24–25
  Agmon–Motzkin–Schoenberg algorithm, 95.178
  All Global Coverings algorithm, 94.191–192
  All Rules algorithm, 94.192–193
  Andrews algorithm, 85.37
  backward algorithm, 49.57, 69; 85.49–54
  basis algorithms, 89.334–349, 383–384
  binary τ-mappings, 89.336, 366

general basis algorithm, 89.337–349, 363–365, 370–374
gray-scale τ-mappings, 89.336–337, 366–374
rank order-statistic filters, 89.335
translation-invariant set mapping, 89.361–366
Bierman algorithm, 85.23–24, 34, 36, 68
block matching algorithms, 97.235–237
Burg algorithm, 84.303
Carlson algorithm, 85.22
characteristics, 87.265–267
Cholesky algorithm, incomplete, 82.72
compression algorithm, 97.193
Cook–Tom algorithm, 94.18
Cooley–Tukey algorithms, 93.17–19, 27–30, 46–52
design, architectural characteristics effects, 87.270–273
discrete-event system:
  evolution algorithm, 90.96–97, 100–102
  Floyd–Warshall algorithm, 90.54–56, 64
  Karp algorithm, 90.61–63
  merging lists, 90.89
  rectify algorithm, 90.105
  resolution algorithm, 90.100–101, 105
division algorithm, 84.115
domain segmentation, 93.304–307
edge-preserving reconstruction algorithms, 97.91–93, 118–129
  extended GNC algorithm, 97.132–136, 171–175
  generalized expectation–maximization algorithm, 97.93, 127–129, 153, 162–166
  graduated nonconvexity algorithm, 97.90, 91, 93, 124–127, 153, 168–175
extended GNC algorithm, 97.132–136, 171–175
EZW algorithm, 97.221, 232

fast computation algorithm:
  of ICT, 88.57–58
  of Walsh transforms, 88.25
Fast Fourier transform (FFT) algorithm, 9.57–58; 86.101, 202, 205, 210; 93.16
  Cooley–Tukey algorithm, 93.17–19, 27–30, 47–52
  Good–Thomas algorithm, 93.19–21
  reduced transform algorithm, 93.16–17, 21–27
fast local labeling algorithm, 90.407, 410–412, 507
Floyd–Warshall algorithm, 90.54–56, 64
forward algorithm, 85.49–50, 54, 69
Gaussian algorithm, 85.37, 43
general basis algorithm, 89.337–349, 363–365, 370–374
general decomposition algorithm, 89.371
generalized expectation/maximization (GEM) algorithm:
  image processing, 97.93, 153
  tomographic reconstruction, 97.162–167
generalized Lloyd algorithm, 84.10
Gerchberg–Papoulis algorithm, 87.14, 16, 20; 95.178–179
Gerchberg–Saxton algorithm (G–S algorithm), 7.220, 224, 240, 243; 13.297–298; 93.110
Gibbs sampler algorithm, 97.119–120, 123
Good–Thomas algorithm, 93.2, 19–21
graduated nonconvexity (GNC) algorithm, 97.90, 91, 93, 124–127, 153, 168–175
group algebra matrix transforms, 94.14, 44
group invariant transform algorithms, 93.1–55
GT–RT algorithm, 93.45–46
hybrid RT/GT algorithm, 93.25–26
image discontinuities, 97.89–91, 108–118

Algorithms, *continued*
  image processing, 90.355, 368–425
  image reconstruction, 97.91, 118–129, 168–175
  intensity gradient algorithm, 93.319–320
  iterative algorithms, 93.110–111, 145, 148, 167
  Joseph algorithm, 85.21
  Jover and Kallath algorithm, 85.39
  Kalman–Bucy algorithm, 85.17–22
  Karp algorithm, 90.61–63
  labeling:
    fast local labeling algorithm, 90.410–412, 507
    local labeling algorithm, 90.405
    log-space algorithm, 90.405
    naive labeling algorithm, 90.398–400
    stack-based algorithm, 90.405, 417
  Levialdi's parallel-shrink algorithm, 90.390–391, 415, 425
  Levinson algorithm, 84.303
  log-space algorithm, 90.405
  low-bit-rate video coding, 97.237–240
  metropolis algorithm, 97.119, 120
  mixed annealing minimization algorithm, 97.122–123, 158
  multigrid algorithm, 82.329–347, 364
  orientation analysis, 93.319–320, 323, 325
  for OSS rules, 89.59–71
  overlapped block matching algorithm, 97.235, 237–252
  parallel image processing, 90.424–425
    Abingdon Cross benchmark, 90.386, 388–389, 425
    fast local labeling algorithm, 90.407, 410–412
    global operations, 90.369–371
    image-template operations, 90.371–382, 417–420
    labeling, binary image components, 90.396–415
    Levialdi's parallel-shrink algorithm, 90.390–396, 415, 425
    local labeling algorithm, 90.400–415
    log-space algorithm, 90.405
    naive labeling algorithm, 90.398–400
    stack-based algorithm, 90.405, 417
  parameter estimation, 94.372–376
  phase retrieval, 93.110–112
    two-dimensional, 93.131–133
  prime factor algorithm, 93.2
  QR algorithm, 85.33–34, 63
  Rauch–Tung–Striebel algorithm, 85.50
  rectify algorithm, 90.105
  reduced transform algorithms, 93.16–17, 21–27, 30–46
  relationship with architectures, 87.261–273
    algorithmic characteristic effects, 87.263–267
    architectural characteristic effects, 87.268–273
    impact on hardware architectures, 87.267–268
  resolution algorithm, 90.100–101, 105
  row–column algorithm, 93.45
  SA-W-LVQ algorithm, 97.220–226
  simulated annealing minimization algorithm, 97.120–122, 155
  Single Global Covering algorithm, 94.188–190, 191
  Single Local Covering algorithm, 94.190–191
  smoothing algorithm, 85.49–50
  square-root algorithm, 85.22–24
  stack-based algorithm, 90.405, 417
  suboptimal algorithms, 97.124–127
  toboggan algorithm, 92.34
  two-grid algorithm, 92.330
  Winograd algorithm, 94.18
Algorithm selection rule, LLVE, 86.121, 123
Algorithm state tree, 86.108, 108–110, 109
Aliasing, 91.244
Alignment:
  aberrations, 13.195, 196
  mirror electron microscopy, 4.215–217
Alkaline phosphatase:
  azo dye technique, 6.186, 187, 188

capture agents, 6.184
colored FRP, 6.186
distribution, 6.172
electron microscopy, 6.173, 174, 184–188
functions, 6.189
Gomori method, 6.172, 173, 184, 191
histochemical reactions, 6.173, 174
histochemistry, 6.172, 173, 174, 180, 186
hydrolysis of ATP, 6.190
incubation, 6.184
localization, 6.184
Menten method, 6.173, 174
optical microscopy, 6.173, 174, 184, 186
resolution, 6.184
specific inhibitors, 6.186
specificity, 6.186
substrates, 6.186
Alkylamines, basic film preparation, use in, 8.126, 128
Allgemeine Elektrizitäts-Gesellschaft (AEG), 10.219; 91.262, 274
electrostatic EM, 10.224–225
EM 5, 96.426–427
EM 6, 96.428
EM 7, 96.433, 438
EM 8–I, 96.438–439
EM 8–II, 96.445, 447
World War II impact, 96.428
Zeiss collaboration, 96.428, 433
All Global Coverings algorithm, 94.191–192
Alloys:
energy selected images, 4.356, 357, 358
loss spectrum of electrons, 4.331, 333, 334, 337, 338–340
Alloy transistor, 91.144
All Rules algorithm, 94.192–193
α-blending, 85.139
α channel, 85.92
α particle, 86.28
Aluminum:
energy selected images, 4.346–350, 353

loss spectrum of electrons, 4.265, 328, 332, 333, 336
as support films, 8.112
Aluminum antimonide, 91.175–176
Aluminum arsenide, 86.49
Aluminum formate support films, 8.123
Aluminum gallium arsenide, 86.44, 48
Aluminum oxide, as substrate material, 83.34, 35, 37
Ambient magnetic field, 83.134
Aminopeptidase, histochemistry, 6.188
Ammonium sulfide, 86.39
A-mode, 84.326
Amoeba, 5.336
Amoeba proteus, holography, 10.191, 192
Amplifier:
history, 91.153–156, 194
lock-in, 1.92, 106
traveling-wave tube amplifier, 91.194
Amplitude contrast, 2.204; 11.27
Amplitude contrast devices:
KA, 6.82–86
theory, 6.81, 82
Amplitude division interferometry, 99.193–194
Amplitude filters, 8.14–22
Amplitude grating, 8.6, 20; 12.52
Amplitude-modulated (AM) signal, A-Psi GDE, 94.340
Amplitude object, 6.50, 51
Amplitude ring, 6.82
Amplitude transfer function, 89.8
Amplitude transmittance, 89.3, 10, 33, 34
confocal systems, 10.22–23
Amplival microscope, 6.105, 108
Analog optical fiber communication, 99.68
Analog recording photoplate, 9.26
Analog system, 91.195
Analog video data, conversion to digital, 4.88
Analysis:
bottom-up, 86.160
edge-based, 86.125, 129–130
optical transforms, 7.55–62
plan-guided, 86.125, 132–133

Analysis, *continued*
  region-based, 86.125, 129–130
  top-down, 86.160
Analysis-by-synthesis speech coding, 82.148, 160, 161
Analysis plan, 86.94
Analytical electron field emission:
  aperture-ring combinations, 8.236–238
  calculation, 8.228–238
  cathode parameters, 8.233–236
  curved or polygonial contoured systems, application to, 8.238
  entire system potential, 8.237
  field strength, absolute value, 8.235
  general description of the model, 8.229–230
  general field, 8.236–238
  illustration of the meaning of the parameters appearing in the theory, 8.232
  logarithmic potential terms, 8.230–232
  vicinity of the apex, 8.232–236
Analytical electron microscopy:
  commercial development in Europe, 96.489–492
  development, 10.257–258
Andesite, 7.37
AND gate, 89.229–231; 90.6–7
Andrews algorithm, 85.37
Anesthesia, effects, 94.290
Angle of illumination, phase object approximation, 7.194
Angles-coded image, 93.239–243, 318, 320
Angular aberration, mirror-bank energy analyzers, 89.403–404
Angular divergence:
  and chromatic aberrations, 13.285–292
  in electron wave optics, 13.288–292
Angular emission field distribution (AED), 8.215–217
Angular intensity, 11.111
Angular measurement, 3.81, 82
Animal intelligence, 94.264, 267–272, 286–287

Anisotropic determinantal equation, 92.112–114
Anisotropic media:
  chiral and chiral–ferrite media, 92.117–132
  computer simulations, 92.133
  electromagnetic properties and field behavior, 92.80–95, 114–117, 132–158, 200
  ferrite media, 92.158–183
  guided-wave structures, 92.183–200
  planar guiding structures, 92.95–132
Anisotropic reaction theorem, 92.119–120
Anisotropy; *See also* Index of anisotropy
  adoptive extremum sharpening filter, 92.44–45
  crystalline magnetic materials, 98.325–326
  in high-temperature superconductivity, 14.29–30
Annular detector, size, in STEM, 7.271
Anode, 83.49, 67
Anode lens:
  electrostatic field emission, 8.242–243
  potential, 8.240
Anomalous diffraction, 11.35
Anomalous transmission (electron), 2.185
Anoptral contrast equipment, 6.66
Anpassung value, 86.178
Antenna, properties of:
  bandwidth, 82.16, 20–22
  input impedance, 82.16, 19–24
  polarization, 82.16, 19–24
  radiation pattern, 82.7, 24–26, 35–40
  radiation resistance, 82.16, 20–23
  resonant length, 82.20–24
Anticontamination devices, Japanese contributions, 96.697–698
Anti-extensive τ-mapping, 89.350–351
Antiferromagnetism, ground-state computing, 89.226–228
Antigens, fluorescence microscopy, 6.117
Antiparallel arc, defined, 90.37
APDs *See* Avalanche photodiodes
Aperture, 5.101, 176, 182, 183, 187

aberration figures, 1.234, 257
  optimum (electron optical), 1.122, 136, 144, 147
Aperture contrast, 4.71–74
Aperture function, 12.27
Aperture-limited wet replication hydration chamber, 8.72–74
Aperture plasmons, 12.268
Apochromat microscope objective, 14.277–278, 295–297
Apple chlorotic leafspot virus:
  concentrated preparation, 6.263
  electron micrograph, 6.270
Application-driven methodologies, 87.260, 263, 285–296
  hybrid parallel architectures, 87.285–291
  model-driven parallel architectures, 87.292–296
Approximately analytical method, 91.3
Approximate maximum likelihood (AML) method, 84.302–309
Approximating function, 82.46, 50, 55, 56, 58, 62, 66, 67, 68
Approximations, 84.244, 247
  Born approximation, 14.220; 95.313
  canonical aberration theory, 91.1–35
  Chebyshev approximation, 90.34–35
  Debye–Huckel approximation, 87.108
  Derjaguin approximation, 87.65–66, 67, 93, 108, 130
  distorted wave approximation, 90.273–275
  distorted wave Born approximation, 90.217, 218, 275, 279
  effective mass, 82.236
  Fraunhofer approximation, 94.202
  Fresnel approximation, 94.202, 203
  general approximation, 90.111–112
  parabolic rigid band approximation, 82.203
  periodic continuation approximation, 86.191
  phase-object approximation (POA), 99.174, 175–176, 184–186
  plasmon pole approximation, 82.205, 210
  "point-probe approximation," 87.136–137
  property approximation, 82.336
  random phase approximation (RPA), 82.205
  stationary-phase approximation, 86.194, 195
  successive approximation quantization, 97.205–216
  successive approximation wavelet lattice vector quantization, 97.191–222
  sudden perturbed approximation, 86.107, 208, 210
  Thomas–Fermi approximation, 82.210
  ultrahigh-order approximation, 91.1–35
  "weak overlap approximation," 87.108
Approximation scaling factor, 97.211
Approximation vector, 97.211
APW, 86.221
Architecture:
  characteristics, 87.269–270
  pipeline, 85.169
  relationship with algorithms, 87.261–273
    algorithmic characteristic effects, 87.263–267
    architectural characteristic effects, 87.268–273
    impact on hardware architectures, 87.267–268
Architecture-driven methodologies, 87.260, 262–263, 273–285
  butterfly multiprocessor architecture, 87.280–281
  chip architecture, 89.217–243
  coarse-grained parallel architectures, 87.279–284
  Cray computers, 85.271–273
  fine-grained parallel architectures, 87.274
  flexibly coupled multiprocessor architectures, 87.290–291
  Harvard architecture, 85.176

Architecture-driven
 methodologies, *continued*
 hierarchical bus architecture,
  87.282–283
 hybrid parallel architectures,
  87.285–291
 interconnectless architecture, 89.219,
  224
 mapping problem, 87.284–285
 massively parallel architecture, 87.273
 model-driven parallel architecture,
  87.292–296
 multiple instruction stream multiple data
  stream (MIMD) architecture,
  85.260–261; 87.262, 269, 272, 280,
  294–295
 optical symbolic substitution (OSS) *See*
  Optical symbolic substitution
 parallel architectures, 87.260
 quantum-coupled architectures,
  89.217–243
 single instruction stream multiple data
  stream (SIMD) architecture,
  87.262, 274, 277–279, 282, 287,
  294–295
 very large instruction word (VLIW)
  architecture, 87.269, 270
Arcing, 83.7
 destructive, 83.67, 80
 self-quenching, 83.65
Arc-set, defined, 90.37
Arcsin function, inversion, 8.40–41
Arc weighting, 90.37
Area, measurement of, 3.67–69
Area microkymography, 7.76, 79
 object display, 7.89
 quantitative evaluation, 7.89
 registration principles and techniques,
  7.78
Area scan:
 combination with linear scan, 6.296
 correction for drift, 6.285
 counting statistics, 6.283
 maximum size, 6.279
 quantitative investigation, 6.277

with ratemeter signal, 6.294
 speed, 6.288
 standard X-ray, 6.288
Argon, melting point, 98.9, 10
Arithmetic:
 combination digital systems, 89.233–234
 Minkowski addition and subtraction,
  89.328
 optical symbolic substitution, 89.59–71
Arithmetic coding, 97.194
Arithmetic–geometric mean divergence,
  91.38
ARMA systems, signal description, 94.322
AR process *See* Autoregressive process
Array computers, 85.25–26
Array processor, 85.15, 26, 45–46, 62
Arrays:
 cone arrays, directional solidifying
  techniques, 83.43–44
 high current emission from, 83.14
 microemitter, 83.54
 microwave power amplifier tubes,
  83.77–78, 80
ART *See* Algebraic reconstruction
Artifacts:
 biological macromolecular interface
  interaction, 8.114–119
 embedding, 2.319, 321, 323
 fixation, 2.254
Artificial intelligence, 94.277
 Turing test, 94.266–272
Ashinuma, Kanichi, 96, 659
Ash minerals, identification of, 3.135, 145,
  152
 in bacteria, 3.126–131, 139
 in mitochondria, 3.118–123, 135
 in tissues, 3.150–152
 in viruses, 3.123–126, 152
Asian–Pacific Electron Microscopy
 Conference:
 first meeting, 96.605–606
 history of meetings, 96.607, 629
Aspergillus terrens, 5.333
Assembly process, ferrite media,
  92.179–181

Assignment, minimax algebra, 90.114
Associativity, fuzzy relations, 89.286
Astigmatism, 1.125, 154, 163, 250; 11.43
  in microscope objective, 14.260–265
Astrophysics:
  delayed-choice experiment, 94.298
  optical astronomy:
    phase retrieval, 93.131
    stellar spreckle interferometry, 93.143–144
  radioastronomy, 91.285–290
    band limited constraint, 7.244
    interferometers, 91.286, 287, 289
Asymptotic technique, 86.261
AT&T, semiconductor history, 91.149, 168, 198
Atomic absorption spectroscopy (AAS), 9.290–292
Atomic beam splitter, ferromagnetic nanotip application, 95.145–149
Atomic force microscope (AFM), 83.87, 88, 89; 87.50
Atomic layer epitaxy (ALE), 86.66; 89.222
Atomic metallic ion emission, 95.94
Atomic sites, 83.61
Atomic structure, 12.74
  crystal-aperture scanning transmission electron microscopy, 93.73–79
Atoming imaging *See* Imaging
ATP:
  histochemical localization, 6.190
  hydrolysis, 6.189, 190
  lead chelate complex, 6.192
  phosphate release, 6.189
  as phosphorylating agent, 6.190
  pyrophosphate release, 6.190
  spontaneous hydrolysis, 6.189
ATPase, 6.174, 204
  capture agent, 6.191
  electron microscopy, 6.191
  histochemical techniques, 6.191–195, 221
  optical microscopy, 6.191
  sensitivity to fixatives, 6.193
  sensitivity to lead, 6.192, 193

  in transport, 6.220
Attention, 94.289–291
Attenuation, 84.320–321, 323, 329, 332
  effect, 84.329
  scattering, 84.321, 324
Attenuation coefficient, estimation, 84.323, 324
Attenuation correction, high frequency, 4.101–102
Attribute significance, rough set theory, 94.182–183
Auger analysis, 11.47
Auger electron spectroscopy (AES), 10.303–304, 305–307; 11.60; 12.93, 108, 110, 112, 113
Auger microscope, 12.93
Auger voltage contrast, 12.148
Austenite–ferrite transition, 10.296
Australia:
  Australian Society for Electron Microscopy, 96.39–40, 46–47
  contributions to electron microscopy, 96.40–42, 50–52
  electron microscopy centers, 96.47–50
  National Committee for Electron Microscopy, 96.42–44, 46
  radioastronomy, 91.287
  research funding sources, 96.39
Austria:
  Austrian Society for Electron Microscopy, 96.55–57
  Glaser, Walter, 96.59–66
Autocorrelation function, 7.108
  axial, 84.331, 334
  lateral, 84.331, 334
  spatial, 84.328
Autocorrelation matrix, 82.121
Autocorrelation method, for linear prediction of speech, 82.123
Auto covariance function, 84.339
Auto-magnification, direct imaging of nucleus, 93.87–89, 104
Automated cervical smear analysis, 9.350
Automated field-tracking system, functions, 6.38

Automated information retrieval, 89.306–307
  fuzzy relations, 89.272, 306–310
Automated microscopy systems, 9.326, 342–346
Automatic gain control (AGC), 99.73
Automatic microscope:
  finding and framing, 5.53–55
  image analysis, 5.60–72
  image display, 5.55–60
  oscillating scanner, 5.50–53
  stage positioning, 5.53
  in white blood cell differential count, 5.50
Automatic programming, 86.98–99, 104, 107
Automation, orientation analysis, 93.320–322
Autoradiography, 2.151; 99.264
  absolute quantitation in, 3.263
  background, 2.158; 3.239, 240
  cross-fire effects, 3.259–263
  of drugs and hormones, 3.226
  efficiency, 3.266–267
  electron microscopical, 3.219
  embedding for, 3.219–228
  emulsion and, 3.260–263
  grain distribution in, 3.243–245
  junctional regions in, 3.243–245, 255
  photographic technique, 3.238, 239
  retention of lipids in, 3.221–228
  retention of macromolecules in, 3.220, 221
  silver grains, 3.243–245
  statistical analysis in, 3.241, 243–263
  with tritium, 2.164
  ultramicrotomy in, 3.228–238
Autoregressive-moving average (ARMA) process, 84.288–296
Autoregressive process:
  experimental results, 84.302–309
  final conditions, 84.299
  initial conditions, 84.297–299
  parameter estimation:
    Box–Jenkins, 84.297–298
    covariance, 84.291
    direct method, 84.288
    exact likelihood, 84.296–297
    finite sample size results, 84.292–293
    forward–backward, 84.298–302
    relation to Jordan algebra, 84.293, 295–296
    transformation method, 84.288–289, 291
  power spectral density, 84.287, 292, 302
  used in example, 84.287
Autoregressive systems, signal description, 94.322, 325
Avalanche buildup time, 99.86
Avalanche multiplication, 99.81–82, 84
Avalanche photodiodes (APDs):
  avalanche multiplication, 99.81–82, 84
  bandwidth, 99.86
  critical device parameters extraction, 99.102–120, 153, 155
  dark current, 99.74, 88, 150–153, 154, 156
  germanium APDs, 99.89–90
  indium phospide-based, 99.142–146, 156
  low-noise and fast-speed heterojunction APDs, 99.94–96
  for multigigabit optical fiber communications, 99.65–156
  multiplication excess noise, 99.84–86
  multiquantum well/superlattice (MQW/SL) photodiodes, 99.94–96
  photocurrent, 99.87
  photogain, 99.74, 120–135, 155
    temperature dependence, 99.135–150
  planar, 99.96–102, 121
  quantum efficiency, 99.87
  rate equations and photogain, 99.82–84
  SAGCM InP/InGaAs APDs, 99.73, 92–94, 99–156
  SAM and SAGM InP/InGaAs APDs, 99.90–92
  silicon APDs, 99.89
  staircase avalanche photodiodes, 99.94–95

Avalanching:
  microplasmas, 10.61
  vacuum microelectronics, 83.21, 29, 31, 34, 85
Average stage-time, 90.59
Averaging identification filters, 94.385
Awareness:
  mind, 94.262–263
  neural network model, 94.289–291, 293
Axial astigmatism, 13.278–284; 86.239, 240, 278
Axial chromatic aberration, correction of, 14.270–277
Axial field computation in magnetic lens, 13.51–54
Axial flux density:
  Glaser model, 13.143
  Gray model, 13.141–142, 144
  Grivet–Lenz model, 13.142, 144
Axial resonance scattering, Bloch waves, 90.302–310
Axiom, 86.144–145
Axiomatic justification, 90.16
Axis conventions, orientation analysis, 93.279–280
Azo dye technique:
  for alkaline phosphatase, 6.186–188
  with aryl thioester azo dye, 6.188
  in electron microscopy, 6.186, 187, 188
  for esterases, 6.187, 197, 198
  with heavy metal diazonium salt, 6.186
  with hexazonium pararosaniline, 6.187, 188
  with indoxyl azo dye, 6.188
  in optical microscopy, 6.186
  with osmiophilic azo dye, 6.186, 187, 188, 198, 214
  with p-mercuricacetoxyaniline diazotate, 6.186

B

Bacillus anthracomorphe, 5.335
Bacillus megaterium, 5.310, 318
Bacillus mesentericus, 5.335
Bacillus mycoides, 5.335
Bacillus proteus vulgaris, 5.318
Bacillus subtilis, 5.318, 319, 335
Background radiation:
  at low concentration levels, 6.287
  estimation, 6.287
  relation to emitter, 6.286
  variation, 6.287
Background subtraction, 92.21
  gray-value tracking, 92.24
  linear regression, 92.22–24
Backpropagation, 94.288
Backscattered electron images, 93.231
Backscattered electrons, 83.205–206, 217
  beam voltage, 83.238, 253
  charging, to avoid, 83.256
  colloidal gold labeling, 83.214–215, 226–227
  cryo-techniques, 83.253–256
  detectors (BED), 83.133, 206, 223–226
  and electrostatic charging, 13.225–226
  low beam voltage, 83.225–226, 238, 253
  in low-voltage inspection, 13.127
  Monte Carlo, electron scattering simulations, 83.218–219
  radiation damage, 83.253–254
  resolution, 83.238, 253
  spectrum, 84.320–321, 324
    intercept, 84.324
    slope, 84.324
  suppression, 83.236
  z contrast images, 83.214–215, 224, 227, 248, 251
Backtracking, 86.98, 125
Backward adaptation, in linear prediction, 82.129
Backward Kalman filter, 85.49–54, 57, 69
Backward orbit, 90.21
Backward predictor, 84.299
Backward recursion, 85.43; 90.19–21, 35
Bacteria, 5.305; 10.312
  chemotaxis, 7.11
  electron microscopy of, 2.242 et seq.
  flagella, 5.300, 301, 302; 96.280, 742

Bacteria, *continued*
  manipulation on tracking microscope, 7.5–8
  movement, 7.5
  radiation damage, 5.335, 336, 343
  rhodopsin, 8.108
  spores:
    mineral content, 3.126–131, 139
    radiation damage, 5.318, 319
    tracking microscope and, 7.1, 11, 12
    ultrastructural studies, Japanese, 96.735, 741–744
    vegetative radiation damage repair, 5.338
    viruses and DNA, 10.311–312
Bacterial viruses, 6.228, 241, 242
Bacteriophage–dye complexes, 8.59
  orientation, 8.60
  sinusoidal intensity of fluorescence, 8.62
Bacteriophages, 5.314
  base plates, 7.303
  DNA arrangement, 8.66
  electron micrograph, 6.228; 7.357
  elementary disc, 7.360, 362
  half-plane image, 7.233
  negative staining, 6.227; 7.288
  optical diffraction pattern, 7.359
  polysheath, 7.366
  protein unit rearrangement, 7.365
  radiation damage, 5.318
  sheaths, three-dimensional structure, 7.338
  specimen preparation, 8.116
  structure, 7.364
  tails, 7.313, 361, 363
    contraction of, 7.361, 364
    three-dimensional structure, 7.295, 356
    three-dimensional reconstruction, 7.359, 368
    three-dimensional structure, 7.295, 356–367
Baker–Campbell–Hausdorff formula, 97.347
Ballistic transport, 83.2, 68; 89.107

Aharonov–Bohm effect-based devices, 89.144–145, 151–152, 166
  energy, 83.18
  motion, 83.18
  space charge effects, quantum mechanical analysis, 89.130–133
Band:
  anisotrophy of, 82.209, 213, 221
  conduction, isotropic, 82.209
  impurity, 82.198, 206, 245
  rigid parabolic, 82.209
  valence, complex structure, 82.213
Bandgap, 82.244
Bandgap semiconductor, tunneling, 99.88
Bandler–Kohout compositions, 89.271–276
  associativity, 89.286
  cuttability, 89.287–288
  fuzzy relations, 89.279–289
Band structure, 82.220, 244
  anisotropy, effect on conductivity tensor, 92.90–93
  electron–impurity interaction and, 82.199
  field electron emission and, 8.218
  fluctuation of impurity concentration and, 82.200
  of silicon and germanium, 82.198
Band tails, 82.206, 245
  formation, 82.200
  Halperin–Lax, 82.234
  Kane, 82.232
  quantum mechanical theory, 82.200, 232
  theory, 82.198
Bandwidth:
  avalanche photodiodes, 99.86, 92–94
  defined, 99.76
Bandwidth estimation, 7.160
Bargmann–Michel–Telegdi equation, 95.375
Barkhausen noise, 87.240
Barnes–Wall lattice, 97.217–218
Barometric height, 85.13–14
Bar patterns, Fourier optics, 10.25–27

Basalt:
  electron probe microanalysis, 6.276
  X-ray color composite, 6.293
  X-ray scan, 6.281, 295, 296
Basic probability assignment, 94.171, 175
Basic probability number, 94.171
Basin stabilization, quantitative particle motion, 98.50–51
Basis algorithms, 89.334–349, 383–384
  binary τ-mappings, 89.336, 366
  general basis algorithm, 89.337–349, 363–365, 370–374
  gray-scale τ-mappings, 89.336–337, 366–374
  rank order-statistic filters, 89.335
  translation-invariant set mapping, 89.361–366
Basis representation, 89.331–333
  dual basis, 89.348–349
  filtering properties, 89.349–358
  gray-scale τ-mapping, 89.368
  transforming, 89.374–383
  translation-invariant set mapping, 89.359–360
Basis restriction MSE, 88.47–48
Bauer's formula, 13.263
Bausch and Lomb "blister" microscope, 5.35
Bausch and Lomb in-cell microscope, 5.35
Bayesian approach, regularization, 97.87–88, 98–104
Bayesian classification, pixels, 97.68
Beam alignment, 83.127
Beam blanking, 83.129
Beam damage, 11.95
Beam deflection system, 11.138–141; 87.192
Beam energy, 83.120, 187
Beam focusing, 83.181, 183, 187
Beam-induced conductivity, 83.234
Beam propagation method (BPM), 93.175
  basic equations, 93.175–178, 186–190
  improved equations, 93.202–207
Beam scanning:
  alternative schemes, 10.70–71
  construction, difficulties, 10.72
  noise and flare, 10.73
  using mirrors, 10.71
  vignetting, 10.71–72
Beam splitter, ferromagnetic nanotip application, 958.145–149
Beam spotsize, 93.182
Beam threshold condition, 90.331–332
Becke line method, 1.65, 66
Beer–Bouger law, 6.150
Belgium:
  Claude, Albert, 96.76–77
  electron microscopy in, 10.221–222
  instrumentation:
    Antwerp facilities, 96.77–78
    distribution, 96.76–77
    history, 96.72–74
  Marton, Ladislaus, 96.67–71, 102, 183
  Society for Electron Microscopy, 96.74–75
Belief function, 94.171
Bell Laboratories, semiconductor history, 91.143, 184, 193–195, 198, 200, 202
Bend extinction contour, 93.60
Bent foil zone axis pattern (ZAP), 93.69, 75, 77
Benzyldimethylalkyammonium chloride, basic protein film preparation, in, 8.125–126
Bertein method, 86.246, 249
Bertram's model of potential distribution, 13.139, 144
Beryllium fluoride support films, 8.123
Bessel function, 13.262–263; 92.109; 94.132–133
Bethe theory:
  angular distribution, 9.97–100
  Compton scattering and, 9.99, 100
  energy distribution of scattered electrons, 9.100–108
  GOS, angular distribution and, 9.97–98
Beyer phase contrast device, 6.88–90
Bhattacharyya's coefficient, 91.43
Bhattacharyya's distance, 91.43, 125

Bhattacharyya's distance, *continued*
   symmetrized Chernoff measure,
     91.128–129
   unified r,s-J-divergence measure,
     91.131–132
Bias, Fisher information, 90.132
Bierman algorithm, 85.23–24, 34, 36, 68
BiFET *See* Bipolar/field-effect cell
Bimetallic wire, biprism, 99.188–190
Bimorph piezosensor, 87.192
Binary components, shrinking, 90.389–396
Binary images, 88.298
   algebra, 90.356
   components:
     area, 90.420
     compactness, 90.421
     computing properties, 90.415–424
     controid, 90.423
     diameter, 90.422–423
     height, 90.421, 423
     labeling, 90.396–415, 417, 507
     moments, 90.423
     perimeter, 90.421
     shrinking, 90.389–396, 407–408
     width, 90.422, 423
   enhancement, 92.64–75
     contour chain processing, 92.70–72
     distance transform, 92.72–75
     polynomial filtering, 92.68–70
     rank-selection filters, 92.65–68
   mathematical morphology, 89.326–327
   porosity analysis, 93.301–308
   representation, 99.38
Binary morphology, 89.326
Binary number system:
   half adder, 89.233
   optical symbolic substitution, 89.58–59
   spin-polarized single-electron logic devices, 89.223–235
Binary relations, fuzzy relations, 89.291–293
Binary space partitioning tree (BSP tree), 85.106, 145–148
Binary τ-mappings, 89.336, 366
Binary threshold model, neurons, 94.276

Binding energy, 82.234
Binding problem, 94.284–285
Binocular image-shearing microscopes, 9.223–242; *See also* Vickers image-shearing module Mark I and Mark II
   binocular equipment, development, 9.227–228
   coincidence setting shear (CSS), 9.240–242
   common path Sagnac interferometer, 9.226–227
   condenser aperture and, 9.238–239
   design, 9.228–230
   dichroic filter, 9.230–231
   Dyson's image-shearing eyepiece, 9.225–226, 229, 230
   edges, 9.236–237
   history, 9.224–227
   Hopkins' image-shearing system, 9.226–227, 231
   image-shearing eyepiece, 9.225–226, 228
   image-shearing measurement procedure, 9.223–224
   low contrast conditions, 9.240
   Michelson system, 9.228–229, 230, 231
   phase shifts, object thickness and, 9.239–240
   precision, 9.236–240
   quasistatistical facility, 9.224
   removal of second image, 9.230
   ringing effect, 9.239
   setting precision, 9.237–238
   tilt threshold, 9.231–232
   transparency of objects and, 9.239
   vibrating mirror system, 9.225
Biological cells, 6.83, 87, 105, 128
Biological macromolecules:
   artifacts produced by interaction with surfaces, 8.114–119
   binding energies, 8.108
   damage to on mounting, 8.113, 115–117
   degree of ionization, 8.109

INDEX

distribution artifacts, 8.114–115
environmental influence on structure, 8.109
main bonds, 8.109
morphological damage during sample preparation, 8.115–117
SiO replication, 8.85
structure preservation, 8.114–119, 122–124
support surfaces for electron microscopy, 8.110, 119–124
surface collapse, 8.116
surface tension damage, 8.115–116
Biological specimens:
  animal intelligence, 94.264, 267–272, 286–287
  Brownian motion, in liquid specimens, 5.339, 340, 343
  cytoplasmic streaming, 5.339
  damage:
    in amorphous state, 5.314
    avoidance, 5.303, 304, 306, 309, 310, 342, 343
    by atomic displacement, 5.308, 309
    by contamination, 5.303, 304
    by electrostatic charge, 5.304, 305
    by ionization, 5.307, 308
    by ionizing radiation, 5.306–309, 313, 335–338, 342
    by sublimation, 5.304
    by temperature, 5.305, 306, 343
    in crystalline state, 5.314
    dosage required, 5.312, 313, 314, 315, 343
    nature, 5.317, 318
  dark-field electron microscopy, 5.300–303, 342, 599
  electrical effects, 94.97
  electron probe microanalysis, 6.282
  homogeneity, 5.339
  irradiation effects, 5.303–309, 311–319, 335–338, 343
  motility, 5.339, 340
  phase contrast microscopy, 6.87
  stained, 6.82, 83, 171
  staining, 5.46, 47, 49, 54, 76, 102
  thickness, 5.338, 339
  viability, 5.335–338
Biology; *See also* Biological specimens; Living cells; Living tissue
  cell structure, 83.245
  histochemistry, 83.214
  LVSEM, 83.242–248, 251, 254–256
  yeast protoplast, 83.244
Bioluminescence, 2.1; 14.122
  development and data of, 2.31–39
  probes, 14.168–171
Biorthogonal functions, Gabor expansion, 97.27–29
Biotite, oligoclase porphyroblast in, 7.37
Biot–Savart law:
  coil field computation, 13.24, 38, 149, 161; 14.58; 82.12, 70, 75
  thin lens elements, 14.73–75
Bipolar diode, 87.217, 230, 232, 244
Bipolar DRAM cells, 86.59–62, 82
Bipolar/field-effect cell (BiFET), 86.69–70
Bipolar memory, 86.59
Bipolar transistors, 82.202; 87.230, 232, 244; 89.136
Bi-potential lens:
  with elliptical defects, tolerance calculations, 13.55–57
  spherical aberration of, 13.48–51
Biprism:
  electron, 12.5, 11, 16, 17; 91.282; 99.173, 176–184
  electrostatic, 99.188–190, 193
  off-axis holography, 89.5, 7, 9–10, 22
Birefringence, 1.74, 78, 81, 88, 95
  crystals, 8.37
  measurement, 1.89
  objects, 6.120
  sign of, 1.92
Bisection method, 82.87
Bismuth, properties, 92.82
Bispectrum, 94.324
Bistability:
  granular electron devices, 89.207

Bistability, *continued*
  quantum-coupled devices, 89.219, 220, 224–225
  resonant tunneling devices, 89.130–133
Bitter pattern technique, magnetic microstructure, 98.328–329
Bivariate entropy, unified r,s, 91.95–96, 98–102
BJT *See* Bipolar memory
Blind deconvolution, 93.144–152
Blind restoration problem, 97.141
Bloch condition, 86.178
Bloch domain wall, 87.177–180; 98.327, 388
Bloch wave, 8.218; 86.176, 178, 180, 186, 187
  axial resonance scattering, 90.302–310
  bound, 90.296
  channelling, 90.293–334
  electron diffraction, 90.226, 229, 240
  one-dimensional, 90.310–317
  perturbation method, 90.251–252
  planar resonance scattering, 90.313–317
  two-dimensional, 90.294–310
Bloch wave-excitation amplitude, 86.187
Block, rough set theory, 94.156
Block adaptation, in linear prediction, 82.129
Block diagrams, 84.247
Block matching algorithms, motion estimation, 97.235–237
Block-parallel methods, image recovery projection, 95.216–217
Block transforms, joint space–frequency representation, 97.11
Blood cells:
  brightfield microscopy, 6.85
  classification, 5.45, 46
  flow of, 1.1; 6.34; 7.90
    velocity, 1.6, 7, 8, 10, 16, 23, 24, 25, 26, 29; 6.39
  KA amplitude contrast microscopy, 6.85
  KFA phase contrast micrograph, 6.74, 75, 76, 77
  KFS phase contrast micrograph, 6.79
  movement:
    by stripe microkymography, 7.96
    speed determination, 7.13, 74
  SiO replication, 8.85
Blood circulation, 6.1, 45
B-lymphocyte fibroblasts, SiO replication, 8.85
B-mode image, 84.325, 327
Boersch, Hans, 96.138–141, 793
Boersch effect, 1.124; 8.220, 256
  biprism, 12.13
  and Coulomb interaction, 13.216–220
  described, 8.189; 96.793
  electron guns, 7.160; 8.138, 178
  low-voltage optics, 13.125
  reducing, 13.222–224
Boersch ray path, 96.139
Bold intersection, fuzzy set theory, 89.263–264
Boltzmann entropy, 90.125, 186
Boltzmann machine (BM), 97.150–151, 153
Boltzmann transport equation, 89.111
Bond distances and angles, calculation of, 88.145
Boolean algebra, fuzzy set theory, 89.258
Born approximation, 14.220; 95.313
Boron, diatomic molecular bond, 98.72–74
Boroscope, 5.16
Bottom-up analysis, 86.160
Bouger law, 6.152
Boundary, 84.199–216
  in adjacency graphs, 84.199, 219–220
  area, 84.200, 216
  Cramer–Rao, 84.286–287, 292, 302, 304–306, 310
  defined, 84.217
  difficulties with, 84.199
  intensity gradient analysis, 93.267–272
  tracking, 84.220
Boundary condition:
  of current vector potential, 82.19, 28
  Dirichlet, 82.11, 12, 14, 18, 27, 44, 46, 51, 52, 54, 58, 62, 63, 67, 70, 85
  of eddy current field, 82.21

INDEX

of electric scalar potential, 82.10, 26, 41
of electric vector potential, 82.42
homogeneous, 82.40, 45
of magnetic scalar potential, 82.12, 28, 42, 80
of magnetic vector potential, 82.16, 26, 41
Neumann, 82.11, 13, 14, 18, 34, 44, 46, 51, 52, 55, 58, 61, 70, 75
of static field, 82.4, 5, 8, 9
three-dimensional circulator model, 98.198–212
impedance boundary condition, 98.288–301
two-dimensional circulator model, 98.90–98
of waveguide and cavity, 82.39
Boundary element method (BEM):
comparison of, 13.169–173
modeling electron optical systems, 13.4
as numerical method, 13.154
Boundary models, 85.96
Boundary value kernels, 92.175–179, 183
Bound Bloch waves, defined, 90.296
Bounded lattice-ordered group, 84.67, 69
radicable, 84.113
Bounded sum, fuzzy set theory, 89.263–264
Bounding, box, 85.213
Box, representation of, 13.29–30
Box dimension, 88.211
Box–Jenkins loglikelihood, 84.297–298
BPM *See* Beam propagation method
Brachet "telemicroscope," 5.36, 37
Bragg condition, 11.90
Bragg diffraction, 84.322
Brain; *See also* Mind
invariants, 94.292–293
mind-body problem, 94.261–262
neural network model, 94.272–273
quantum neural computing, 94.266–310
as quantum system, 94.260
reductionist approach, 94.262–263, 285
scripts, 94.290–291
somatosensory cortex, 94.291

split-brain research, 94.261, 263
vision, 94.273, 283–284, 302
Breakdown, 87.211, 217, 236, 240, 245
Breakdown voltage, temperature dependence, 99.135–146
Break point, 84.244, 248
Bright field microscopy, 6.82, 83, 84, 85, 86, 87, 90, 116; 94.220, 252
complementary half-plane objective apertures, 7.251–252
phase problem and, 7.227, 228, 239–242, 247–249
tilted illumination, 7.249
transmission electron microscopy, 90.211
Brightness:
electron beam, 8.139, 248–251
beam cross-section, as a function of, 8.173
beam current and, 8.181
calculation, 8.166, 171
cathode diameter and, 8.181–182, 183
cathode temperature and, 8.179
cathode wehnelt and cutoff voltage difference, as a function of, 8.181
curves, maxima in, 8.184–185
experimental determination, 8.177–179
grid bias, dependence on, 8.180
maximum elucidation, light optical model, 8.185–186
measured end values, 8.179–181
measurement techniques, 8.168–177, 182–183
measuring aperture, permissible values, 8.170
scattering effect on, 8.185–186
single-crystal cathode tip orientation, influence of, 8.183–186
theoretical value, 8.168, 170, 174, 179–182
thermionic cathode emission, for, 8.166–168
electron sources, 83.221–222, 229
field electron emission, 8.248–251

Brightness, *continued*
  two-aperture measurement, 8.174–175, 179–183
Brightness contrast, 6.51
Brillouin effect, 14.122
British Thompson Houston Company, 91.267, 268
Broadcasting, 91.207–209
Brome mosaic virus, 6.232–265
Bronze, particle size analysis, 5.156–159
Browder's admissible control, image recovery projection, 95.209–211
Brownian motion, in liquid biological specimens, 5.339, 340, 343
BSP tree *See* Binary space partitioning tree
BTE, 89.111
Bubble chamber recording, 10.206–207
Bubbles, quantitative particle motion, 98.44–61
Buffer:
  depth-coordinate, 85.137, 147
  frame, cubic, 85.163
Bugstore, 91.242
Buildup, 83.17
Build-up field emission cathodes, 8.253, 257
Bulirsch–Stoer–Gragg formula, 13.202
Bulk probes, properties, 87.138–148
Bulky superconductors, critical current density in, 14.36–38
Bunsen–Roscoe law, photomicrography, 4.389
Burg algorithm, 84.303
Burg method, 84.302–309
Burstein effect, 10.69; 82.201
Burst (popcorn or telegraph) noise, 83.58
Burton, Eli, 96.79–82, 87
Butterfly multiprocessor architecture, 87.280–281
Buttiker theory, 91.215, 216, 219, 223–224, 227

C

C (computer language), data structures for image processing, 88.63–108
  design, 88.67–81
  image representation, 88.65–66
  implementation, 88.81–108
  object-oriented systems, 88.68–69
C++ (computer language), data structures for image processing, 88.70, 77, 92
Cache memory, 86.4
CAD *See* Computer-aided design
Cadmium sulfide photoresistors, 4.396
Calculus, 89.266–267
  classical relational, 89.267–276
  fuzzy relational, 89.266, 276–289
  rough set theory, 94.151–194
Calibration transforms, optical diffraction analysis, 7.27
Calligraphic systems, 91.235–236, 237, 242
CAM *See* Content-addressable memory
Cambridge Instruments:
  computer control of microscopes, 96.510–511
  electron microprobe production, 96.464–466
  scanning electron microscope production, 96.468–469, 505–506, 509
Cameca:
  Camebax instrument, 96.504
  electron microprobe production, 96.464–465, 519
  MEB 07, 96.470
  niche marketing, 96.532–533
Cameras:
  electronic, digital image processing, 4.145–149
  microkymography attachments, 7.85
  mirror electron microscopy, 4.214
  mirrors in, 89.460–464
  photomicrography, 4.387–389
  television, 4.88–105
  tubes, 9.39–44
Camscan:
  electron microscope development, 96.469–470, 503–504
  niche marketing, 96.532–533

Canada:
  Burton, Eli, 96.79–82
  Burton Society of Electron Microscopy, 96.87
  electron microscopy, 96.85–86
  Hall, Cecil, 96.81–82
  Microscopical Society of Canada, 96.88, 90, 91
  transmission electron microscope prototype, 96.82–85
Canberra distance, 92.35
Cancellation, visual perception, 84.138, 141
Cancellation error, 82.13
Cancer cells:
  carcinoma cells, 10.312–313
  wet replication, 8.99
Canonical aberrations, 97.381–383
  eikonal functions, 91.1–35
  ultrahigh order, 97.360–406
Canonical coordinates, 84.147–149
  rotations and dilations, 84.152
  smooth deformations, 84.154
  translations, 84.152
Canonical equation, 91.13
Canonical expansion, of eikonals, 91.16–28, 34
Canonical isomorphism, finite abelian groups, 93.8–9
Capacitance, interconnect capacitance, 89.209
Capacitance detection system, 87.191–192
Capacitor, 87.240
Capillary action, 87.52–53
Capillary forces, 87.119–127
  critical probe–substrate separation, 87.124
  force-versus-distance curves, 87.121
  as function of probe–substrate separation, 87.122–123, 125–126
  hysteresis effect, 87.123
  Kelvin equation, 87.120
  Kelvin radius, 87.120–121
  long-range, 87.125–126
  meniscus volume, 87.124

CAPSEM *See* Committee of Asia–Pacific Societies for Electron Microscopy
Carbocyanines, fluorescent probe, 14.152
Carbohydrates, fixation, 2.260, 267
Carbon:
  amorphous, 5.308
  diatomic molecular bond, 98.72–74
Carbon dioxide, fluid bubbles in water, 98.44–61
Carbon layer, 1.155, 156, 160
Carbon support films, 8.110, 112, 123
  basic, 8.126
  contamination in the STEM, 8.128–129
  glow discharge treatment, 8.129
  images, 7.161
  thin, focus series, 7.202
Carcinogens, detection, 10.310–311
Cardiac pulsation, 1.4, 6, 27, 30
Cardinality of set, 84.73
Carlson algorithm, 85.22
Carrier frequency, electron off-axis holography, 89.22
Carry-free addition, 89.59, 61
CARS *See* Coherent anti-Stokes Raman spectroscopy
Cartesian ACC, 84.209
Cartesian product, fuzzy set theory, 89.264
Cartography, 84.250–254
Cascaded function call, 88.77
Cascaded τ-mapping, 89.345–348, 363–366, 381–383
Castaing and Henry's analyzer, 10.320
Cat:
  gingival microcirculation, 6.11, 12
  heart:
    atrial blood flow, 6.26, 27
    atrial microcirculation, 6.10
    frequency distribution in motion, 6.39
    power spectrum, 6.39
    vertical motion, 6.39
  mesentery:
    clamp, 6.15, 38
    microcirculation, 6.10, 11
  salivary gland, phase contrast microscopy, 6.93

Catalase, 220; 5.195, 299, 310, 316, 317, 329; 6.211, 255; 8.115
  crystalline, 5.316
  diffraction image plane method and, 7.223
  electron diffraction pattern, 7.299
  electron micrograph, 7.298, 299
  freeze-dry specimen preparation, 8.116
  from erythrocytes, three-dimensional structure, 7.341
  GS algorithm and, 7.222
  image enhancement, 4.105–112
  image reconstruction, 7.216
  micrograph analysis, 7.213
  negatively stained, 7.215
  phase problem solution, 7.220
  signal-to-noise ratio, 7.258
  structure, 3.196–200
  three-dimensional reconstruction, 7.339, 340, 344
  three-dimensional structure, 7.332–335, 340–344
  unstained, electron diffraction, 7.302
Catalysis, wet replication study of, 8.99
Cathode, 2.2, 3, 14; 83.7, 11
  barium oxide coated, 83.27, 28
  cold, 83.6, 79
  displays, use in, 83.79
  fabrication, 83.36–38
  failure modes, 83.62
  Fowler–Nordheim plots, 83.47, 50, 61
  microwave power amplifier tubes, 83.77
  Orthicon, 2.9
  spectral sensitivity of, 2.3, 4, 9
  Spindt cathode, 83.47, 50, 61, 65
  thermal/field emission, 83.33
  thermionic, 83.7, 28, 29, 75
  tubules as, 83.34
  VFET, 83.72
Cathode-coaxial anode, electrostatic field, analytical models, 8.223–224
Cathode orientation effects, electron gun, 8.199–200
Cathode ray oscillographs, 10.218–221

Cathode ray tube (CRT), 83.6, 77–80, 204; 91.231–233, 235–236
  computers, 91.235–236
  flat screen, 91.236
  monoscope, 91.240
  projection displays, 91.253
  rear-port tube, 91.236
  storage tubes, 91.237–238
  vacuum fluorescent displays, 91.236
Cathode space electrostatic field equipotential lines, 8.247
Cathodoluminescence (CL), 11.77–86; 83.81, 205, 214
Causality, 99.6
Causal residual, 84.17
Cavity, 82.4, 38, 39, 40, 45, 61, 92
Cayley–Hamiltonian theorem, 90.118–119; 92.102, 103
CBED *See* Convergent-beam electron diffraction
CCD *See* Charge-coupled devices
CCITT *See* International Telephone and Telegraph Consultative Committee
CCITT ADPCM, 82.153
CDM *See* Charged density method
CE *See* Constitutive equation
Cell death, neuronal, 94.287
Cell refractometer, 1.57, 59, 62, 64–66, 70, 73
Cells:
  abstract, 84.202
  cations, probing of with fluorescence microspectroscopy, 14.160–167
  complex, abstract (ACC), 84.202–208
  connected, 84.208
  defined, 84.202
  k-dimensional, 84.202
  components, identification, 2.325 et seq.
  cytoskeleton, 10.316–318
  fluorescence microspectroscopy, 14.150–152
  d-dimensional, 84.202
  histogram, 6.149
  identification by enzyme content, 6.176
  list, 84.224–228, 247, 250, 254

membranes, 5.341
movement, 6.77
radiation damage, 5.342
metabolism, non-invasive probing with fluorescence, microspectroscopy, 14.134–137
organelles:
  identification by enzyme content, 6.176
  probing of with fluorescence microspectroscopy, 14.137–160
  surface imaging, 8.99; 10.312–319
CELLSCAN/GLOPR system, 5.44, 45
  applications, 5.72–77
  for chromosome counts, 5.75
  colored illumination, 5.56, 60
  contourograph, 5.57
  cytological data sheet production, 5.73
  display, 5.55, 56, 57
  for epithelial cell analysis, 5.76
  finding and framing, 5.53, 54, 55
  fully automatic, 5.72
  image processing, 5.56, 60–72
  image recording, 5.56
  layout, 5.50
  microscope stage positioning, 5.53
  oscillating mirror scanner, 5.50–53
  speed, 5.76, 77
  for white blood cell differential, count, 5.72, 73
Cellular array machines, 84.64; 90.355
Cellular automata, 89.217, 220, 224, 229, 240–241
Cellular logic image processor (CLIP series), 87.273, 274–275
Cellulose fiber, 6.120
Cell walls, 5.300, 301, 302
Center of potential, 86.202
Center-weight median filtering, 92.17
Centrable rotation apparatus, 1.54, 55
Centrifugation, support film sample preparation in, 8.121
Centroid condition, 82.138
Centrosymmetric condition, 86.180

Ceramics, use of image analyzing computer, 4.380
Cermet, 83.26
Certainty, rough set theory, 94.166
Cervical smear, 4.362; 5.76, 77
CESEM *See* Committee of European Societies for Electron Microscopy
Cesium, 11.108; 83.17, 28, 29, 30, 34
C evaporant gas, wet replication, in, 8.77
Chadwick–Helmuth flash system, 6.34
Chain code, 85.103
Chaining, 85.25
Chalnicon, 9.343
Change effect transistor, 89.204
Chaotic dynamics, 94.288–289
Character basis, 93.8
Character group, 93.6–9
Characteristic function, 84.7, 75; 89.366
Characteristic mapping:
  classical relational calculus, 89.275–276, 282
  fuzzy relations, 89.286–287
Characteristic maxpolynomial, 90.118
Characteristic state, 97.410
Characteristic surface, spacetime algebra, 95.310–311
Charge, 82.5
Charge carrier, 83.73
Charge-coupled devices (CCD), 9.44, 45–46, 58, 60; 89.221, 244
  detection at high energies, 9.52, 143, 144
  deterioration with prolonged irradiation, 9.45, 46, 145
  normalization mode and, 9.45, 46
  picture acquisition and, 9.326
  solid-state image converters and, 9.44, 45
  tandem optics, 9.45
  TV-image intensifier design and, 9.31
Charged density method (CDM):
  comparison of, 13.172–173
  modeling electron optical systems, 13.4
  as numerical method, 13.154, 162–165
Charged dielectric spheres, electron holograms, 99.174, 207–216

Charge density, 82.2; 86.258, 259
  surface:
    electric, 82.5, 6
    magnetic, 82.8
  weak phase approximation and, 7.267
Charged microtips, electron holography, 99.229–235
Charged-particle wave optics, 97.257–259, 336–339; *See also* Electron wave optics
  Dirac equation, Foldy–Wouthuysen representation, 97.267–269, 322, 341–347
  focusing, 89.392
  Green's function:
    for nonrelativistic free particle, 97.280, 350–351
    for time-dependent quadratic Hamiltonian, 97.351–355
  Klein–Gordon equation, 97.259, 276, 337, 338
    Feshbach–Villars form, 97.263, 322, 339–341
  Magnus formula, 97.347–349
  matrix element of rotation operator, 97.351
  scalar theory, 97.316–317
    axially symmetric electrostatic lenses, 97.320–321
    axially symmetric magnetic lenses, 97.282–316
    electrostatic quadrupole lenses, 97.321–322
    free propagation, 97.279–282
    general formalism, 97.259–279
    magnetic quadrupole lenses, 97.317–320
  spinor theory, 97.258
    axially symmetric magnetic lenses, 97.333–335
    free propagation, 97.330–332
    general formalism, 97.322–330
    magnetic quadrupole lenses, 97.226
  trajectories, 89.393–399
Charge simulation method (CSM) of
  numerical analysis, 13.129, 152, 165–169
  comparison of, 13.172–173
Chebyshev approximation, 90.34–35
Chebyshev distance, 90.23
Chemiluminescence, 14.122
Chernoff measure:
  generalized, 91.125
  symmetrized, 91.126–129
Chicken:
  embryonic heart contractions, stripe kymography and, 7.96
  erythrocytes, chromatin, 8.95, 97, 100–101, 103
Chi function, 94.327
Childs–Langmuir equation, 83.49, 51
Childs–Langmuir law, 11.105
China:
  cultural revolution impact on science, 96.819–823
  electron microscopes:
    early history, 96.805–813
    electron opticians, listing, 96.846–847
    first prototype, 96.805–808
    manufacturing activities, 96.843–846
    market, 96.840–841, 843
    XD-100, 96.810–812
  Nanjing Jiangnan Optical Factory, 96.834–838
  Scientific Instrument Factory (KYKY):
    AMRAY union, 96.831–833
    DX-2, 96.817–819
    DX-3, 96.825–828
    DX-4, 96.828–830
    DX-5, 96.828, 830–831
    electrolytic tank production, 96.816
    first electron emission microscope, 96.815–816
    impact of politics, 96.814–817, 819–825
    KYKY 2000, 96.833
    KYKY 3000, 96.833–834
    1000B SEM, 96.833
    origins, 96.814
    sales, 96.841

transmission electron microscopes, 96.838–839
Chinese remainder theorem, 93.4–6
Chip architecture, 89.217–243
Chip density, 87.270–271
Chiral–ferrite media, 92.117, 132
  constitutive relations, 92.119
  dispersion relations, 92.126
  dyadic Green's function, 92.123
Chiral media, 92.117, 132
  constitutive relations, 92.118–119
  dispersion relations, 92.125–129
  electric field polarization, 92.130–132
  vector Helmholtz equations, 92.122–125
Chlamydomonas:
  negative staining reaction, 6.234
  tracking microscopy, 7.9
Chlorophyll, structure, 10.309–310
Cholesky algorithm, incomplete, 82.72
Cholesky decomposition, 85.23, 27
Cholinesterase, 6.174, 195, 197, 209
  acetylthiocholine technique, 6.200–204
  azo dye technique, 6.197
  distribution, 6.207
  electron microscopy, 6.200, 202, 203
  fixation, 6.201
  gold capture agent, 6.203
  histochemistry, 6.196
  localization, 6.202, 203
  optical microscopy, 6.201, 202
  specificity, 6.200
  staining, 6.201
  substrate, 6.196, 198
  thiolacetic acid technique, 6.199
Cholinesterase technique, 6.176
Chopped beam, in one-dimensional scanning, 10.60
Christoffel symbols, 84.190
Chromatic aberration, 2.189, 192, 193, 194; 5.165, 181, 193, 298
  and angular divergence of e-beams, 13.285–292
  at low-beam voltage, 83.221–222
  of C line, 14.318–320
  correction of, 14.270–281

achromat, 14.277–278
apochromat, 14.277–278
axial, 14.270–277
lateral, 14.278–281
secondary spectrum, 14.276–277, 309–320
two colors, 14.273–276
electron lens, 83.207, 221–222, 229
electron optics, 1.125, 132, 267
ion probe microscopy, 11.120, 125, 138
latent:
  correction, 14.278–281
  monochromatic, 14.288–289
long working distance objective, 14.325–326
LVSEM, 83.207, 229
microscope objective, 14.267–270
mirror-bank energy analyzers, 89.404–408
  equations, 89.397–399
  transaxial mirrors, 89.448–451
objective lens, 7.260
off-axis holography, 12.68
plane parallel glass plate, 14.312–316
reflection electron microscope, 11.68
scanning transmission electron microscopy, 93.87
of second rank, 13.194–195, 196
thin film stack pattern, 14.96–97
two-color, 14.273–276
wave function and, 13.285–288
Chromatic partial coherence, 89.25–26
Chromatin:
  DNA arrangement in, 8.52
  microsurface spreading, 8.93–97
  negative staining, 7.288
  wet replication investigation, 8.52–53, 85, 93–97, 99
Chromium dioxide, fine magnetic particles, 98.415–422
Chromium tribromide, 5.293
Chromosome analysis, 2.136, 144
Chromosome–dye complexes:
  dissociability determination, 8.58

Chromosome–dye complexes, *continued*
  orientation, 8.60–63
  preparation, 8.58–60
Chromosomes, 5.75
  dipoles orientation distribution, 8.70
  DNA arrangement in, 8.52
  electron microscopy, 7.288
  human, 5.159, 160; 6.160
  microsurface spreading, 8.95–97
  microtubular bridges in, 8.95–96
  mitotic cells, isolation from, 8.93
  orientation, 8.66
  polarized fluorescence microscopy, 8.53–63
  subunit surface topography, 8.95
  surface tensional stretching, 8.95, 98–99
  transmission images, 8.95, 98–99
  wet replication examination, 8.52–53, 85, 93–97, 99
Circles, transforms, 7.26
Circuit parameters:
  ferrite media, 92.181–183
  two-dimensional microstrip circulator model, 98.108–116
Circuits, solid state, imaging, 4.248
Circular harmonic expansion, 84.144
Circular mean, orientation analysis, 93.281
Circular standard deviation, orientation analysis, 93.282
Circular variance, orientation analysis, 93.282
Circulators:
  three-dimensional theory, 98.127–129, 301–303, 316–317
    boundary conditions, 98.198–212, 288–301
    characteristic equation through rectangular coordinate formulation, 98.151–170
    diagonalization of governing equation, 98.139–151
    doubly ordered cavity, 98.283–287
    dyadic recursive Green's function, 98.128, 219–238, 316

  field equations, 98.129–139
  impedance boundary condition, 98.288–301
  limiting aspects, 98.246–260
  metallic losses, 98.195–198
  nonexistence of TE, TM, and TEM modes, 98.128, 174–176
  Nth annulus–outer region interface, 98.212–218, 234–238
  radially ordered circulator, 98.260–283
  scattering parameters, 98.238–245, 302
  three-dimensional fields, 98.176–187
  three-port circulator, 98.238–245
  transverse fields, 98.170–174
  z-field dependence, 98.188–195
  z-ordered layers, 98.260–283
  two-dimensional model, 98.80–81, 127, 316
    boundary conditions, 98.90–98
    circuit parameters, 98.108–116
    cylindrical coordinates, 98.83–86
    dyadic recursive Green's functions, 98.79–81, 81–82, 98–108, 121–127, 316
    governing Helmholtz wave equation, 98.86–87
    limiting aspects, 98.121–127
    numerical results, 98.303–315
    scattering parameters, 98.117–120
    three-port circulator, 98.117–120
    two-dimensional fields, 98.87–90
Citrus tristeza virus, concentrated preparation, 6.263
Classical electrodynamics:
  information approach, 90.149–153
  Maxwell's equations, 90.186–187
  from vector equation, 90.196–198
Classical phase plane, 94.348
Clathrates, 8.109
Clausius–Mossotti equation, 87.115
Clifford algebra *See* Spacetime algebra
Clifford–Hammersley theorem, 97.90
C line, chromatic aberration of, 14.318–320

CLIP7, 87.275–276
CLIP7A, 87.276
Cliques, neighborhood system, 97.105–106
Closed-loop codebook design, 82.174
Closed-loop pitch prediction, 82.167
Closed subcomplex, 84.205
Closest encounter techniques in Coulomb interaction, 13.218
Closing:
  function, 99.9–10
  mathematical morphology, 89.329–330, 336, 345, 372–374
  multiscale closing-opening, 99.22–29
Closure, 84.207
  math, fuzzy relations, 89.293–294
C-matrix transform, 88.41
CMU Warp, 87.278–279
Coating, metal, 83.219, 233, 236–239
  conductivity, 83.237, 239
  decoration, 83.233
  double-layer coating, Pt-C, 83.247–248
  ion-beam sputtering, 83.239, 245, 247
  Penning sputtering, 83.236
  Pt-Ir-carbon, 83.239
  structure, 83.238
  thickness, 83.236–237, 247
  topographic z contrast, 83.236
Coaxial cable, 91.193–194
Coaxial projection, three-dimensional reconstruction by, 7.305
Cobalt, magnetic thin films, 98.389–397
Cobalt/copper, magnetic multilayers, 98.406–408
Cobalt crystal, mirror electron microscopy image, 4.257
Cobalt foil, 5.257
Cobalt/palladium, magnetic multilayers, 98.402–406
Codebooks, 82.138
  lattice codebooks, 97.218–220, 227–230
  multiresolution codebooks, 97.202
  regular lattices, 97.218–220
  successive approximation quantization, 97.214–216
  training, 82.173

Code excited linear prediction, 82.173
Coding *See* Image coding
Coefficient selection function, 94.45
Cognitive function:
  animal intelligence, 94.264, 267–269
  scripts, 94.290–291
  Turing test, 94.266–272
Cognitive function reflection, 94.294
Coherence, 2.45, 46, 50; 12.31, 43, 45, 46; 85.85, 92
  edge, 85.158
  in electron optics, 7.101–184
  of electron source, resolution and, 7.259
  face, 85.158
  instrumental aspects, 7.141–171
  scanline, 85.158
  spatial, 85.106, 222
Coherence functions, propagation through optical systems, 7.112
Coherence ratio, 7.141
Coherence theory, classical, 7.105–116
Coherent anti-Stokes Raman spectroscopy (CARS), 10.84–85
Coherent detector, confocal imaging, 10.20
Coherent illumination:
  image formation, 8.3
  phase object approximation and, 7.193
  scanning optical microscopy, 10.22
  transfer theory and, 7.103
Coherent imaging, through turbulence, 93.152–166
Coherent node renumbering, 90.52
Coherent optical system, image processing, 4.129
Coherent transfer function, conventional and confocal systems, 10.26
Cold-field electron emission, 8.255; 9.191
Collagen, band patterns in negative staining, 6.233
Collimators, mirrors, 89.460
Collodion, 5.298, 299
Color, schlieren systems, 8.13–14
Color contrast, 6.51
Coma:
  as aberration, 13.199

Coma, *continued*
  in electron optical transfer function, 13.280–284
  in microscope objective, 14.266–267
Combination function, 86.128–129, 133
Combined aberrations, 91.30–33, 34; 97.360–361
Committee of Asia–Pacific Societies for Electron Microscopy (CAPSEM):
  conferences, 96.31–32, 597, 629
  international federation, 96.629–631
  origin, 96.31
Committee of European Societies for Electron Microscopy (CESEM), 96.28–29, 29, 31
Common dyadic symmetry, 88.17
Communications *See* Telecommunications
Commutator, 84.139, 150, 186
Compensated samples, 82.251
Compensation ratio, 82.230
Compensator, 1.79, 85, 86, 88
  phase shift, 1.97
  types of, 1.79
Complementarity, 94.262, 263, 295, 296
Complementary area, 6.52
Complementary half-plane aperture method:
  bright field microscopy, 7.251, 252
  phase determination and, 7.273
  phase problem solution and, 7.228–238
Complete flux expulsion model, 87.186–187
Complete lattice, 84.63
Complex degree of coherence, 7.109, 111, 112
Complexity statement, defined, 90.13–14
Complex modes, quantum mechanics, 90.156–157
Complex potential, 86.184, 185, 186
Complex spatial filters, 8.20–22
Complex spectrogram, conjoint image representation, 97.9–10
Composite enhancement filter, 92.16–17
Compounding, spatial, 84.341, 342
Compression *See* Image compression
Compression algorithm, 97.193

Compton scattering, 14.122
Compton wavelength, 97.410
Computation:
  combination digital systems, 89.233–234
  optical symbolic architecture, 89.54–91
  paradigm, 94.261, 294
  spin-polarized single-electron chips, 89.224–225
  reading and writing, 89.235–237
Computational complexity, 82.332
Computed radiography, image plate with, 99.242, 263–265
Computed sonography, 84.344
Computed tomography:
  convex set theoretic image recovery, 95.177–178, 182–183
  image formation, 97.59–60
Computer-aided design (CAD), 82.1–4
  in modeling electron optical systems, 13.3
Computer-controlled switching, 91.164, 201–204
Computer optimization, 83.145–152
  "complex" method, 83.146–149
  weight complex method, 83.149–150
Computer programming *See* Programming
Computers:
  array computers, 85.25–26
  digital image processing, 4.153–158
  frame stores, 91.242, 244
  graphic displays, 91.234–236, 237, 244
  high resolution by, 4.113–119
  image analyzing, 4.361–383
  image enhancement, 4.87–102
  image processing, 13.245–246
  massively parallel computers, 90.364
  optical reconstruction by, 7.256
  parallel image processing on, 90.353–426
  programming *See* Programming
  quantum neural computing, 94.260–310
  radioastronomy, 91.287
  stereo images, 91.254
  Turing test for intelligence, 94.272

vector computers, 85.25
virtual reality, 91.255
Computer simulations:
　anisotropic media, 92.133
　crystal-aperture scanning transmission electron microscopy, 93.66–73
　electrical microfield contrast simulation, 94.135–140
　EMS image structures, 94.135–140
　object reconstruction, 93.127–130, 133–134, 162–166
　parameter estimation, 94.375–376
Computer vision, 86.84
Computing models, 94.260–264, 283–284
Concave functions, 91.65–68
Concavity:
　maxpolynomials, 90.107
　Shannon's entropy, 91.38
Concentration mapping technique, 6.297
Condenser-objective lens, 6.21, 22; 10.234–236; 11.7; *See also* Single-field condenser-objective
Condensers:
　cylindrical, 86.246, 247, 250, 256, 261, 274
　toroidal, 86.243, 245, 261, 278
Condiant of a function, 94.332
Conditional entropy, unified (r, s), 91.95–107, 121–123
Conditional information, 90.151–153, 185
Conditional synchronization, 85.266
Conductance:
　quantum conductance, 89.113–117
　universal conductance fluctuations, 91.216–218
Conduction band, 83.2, 7–10, 20, 21, 24, 30
Conduction cooling, 83.63, 64
Conductivity, 82.3, 23, 34, 38, 39
Conductivity tensor, 92.83–85, 90–93
Cones:
　arrays, 83.43
　representation of, 13.29–30
　vacuum microelectronics, 83.14, 26, 36–38, 45, 56, 65, 72

Confined systems, 89.96
Confocal laser scanning microscope, 85.206
Confocal lens, 11.15, 65
Confocal microscopy, 10.30–33; 83.215; 93.283, 325
　amplitude transmittance, 10.22
　annular pupils, 10.20–22, 27, 32
　axial scanning, 10.44, 45
　circular pupils, 10.21–22, 27, 32
　compared with conventional, 10.30–33
　confocal imaging, 10.14–16
　image intensity, 10.20
　multiwave three-dimensional imaging, 14.184–189
　optical sectioning, 10.34–49
　point object, 10.35, 42
　Raman, 14.198–199
　reflection system, 10.23, 24
　　defocused transfer function, 10.27–29
　SAW microscopy, 11.165
　spherical aberration, 10.30
　stereoscopic image pair, 10.47–49
　straight edge response, 10.24
　theory of imaging, 10.16–34
　in tomography microscopy, 14.217
　transmission cross-coefficient, 10.30
Confocal Raman microspectroscopy, 14.196–199
Conformal mapping, 86.227, 246, 247, 269, 274
Conjoint image representation, 97.2–4, 5, 19–37
Conjugacy:
　in image algebra, 84.74, 78
　in semi-lattice, 84.70
Conjugate area, 6.52, 58
Conjugate gradients, 82.72, 77, 81, 348
Conjugation:
　inverting inequalities, 90.30–31
　matrix, 90.4
　products, 90.24–25
　scalars, 90.23–24
Conjunction, fuzzy sets, 89.261, 297–298
Connected complex, 84.208

Connected components, minimax algebra, 90.50
Connected graphs, 90.48–51
Connectedness relation, 84.207
Connectionism, 94.298–299
Connection machine, 87.274–275
Connectivity, 90.45–51
 of complexes, 84.207–208
 paradox, 84.198, 212
  resolution, 84.212–216
 strong connectivity, 90.48–49
Consciousness:
 anesthesia, effects, 94.290
 mind, 94.261, 263
 neural network model, 94.289–291, 293
 as recursive phenomenon, 94.293
 Vedic cognitive science, 94.264–265, 267
Conservant, 94.328, 331
Consistency ratio:
 domain segmentation, 93.298
 orientation analysis, 93.282, 287, 292–293
Consistent labelling, 84.232
Constitutive equation (CE), space-variant image restoration, 99.305–308
Constitutive relations:
 chiral media, 92.118–119
 electromagnetism, 82.2, 3, 5, 7, 10, 12, 16, 19, 23, 26, 27, 38, 41, 42
 gyroelectric–gyromagnetic variational analysis, 92.136
 linear media, 92.135
Constitutive tensor, 92.114
Constraint transformation rule, LLVE, 86.121, 123
Constructability, 94.325
Constructive models, 85.96
 solid geometry, 85.105, 205
Contact angle, 8.111, 112
Contacts, between metals, electrical noise, 87.234
Contact value theorem, 87.103, 108
Contamination, 1.155, 159, 160; 83.15
 at low-beam voltage, 83.221, 239, 240
 specimen borne, 83.238
 vacuum, 83.213, 237, 239, 240
Content-addressable memory (CAM), 86.40; 89.70–72, 88, 90
Contextual region, 92.36
Continuity property, 99.4, 6
Continuous additive and multiplicative maximum, extension, 84.81
Continuous representation, 85.98
Continuous signals, exact Gabor expansion, 97.23–30
Contour chain processing, binary image enhancement, 92.70–72
Contour length measurements, 8.127
Contour maps, numerical simulation, 99.214–216, 227
Contourograph, 5.57, 58
Contour perception, 84.138; 85.103
Contour preservation, 84.343
Contour properties, signal description, 94.318
Contour stack, 85.103
Contraidentity matrix, 84.276
Contrapositivity, 89.277, 278
Contrast; *See also* Image contrast
 of an emulsion, 1.187
 aperture, 4.71–74
 by phase shift, 4.46–67, 81
  axial illumination, 4.46–64
  oblique illumination, 4.65–67
 calculation, electron mirror system, 94.108–144
 curves for interference microscopy, 1.85
 electron, 2.204, 211
 electron image, 1.129, 131, 134
 electron microscopy of biological macromolecules, 7.287
 of emulsions, table of, 1.191
 enhancement:
  adaptive contrast enhancement, 92.34–36
  extremum sharpening, 92.32–34
  imaging plate, 99.285
  inverse contrast ratio mapping, 92.30–31

local range stretching, 92.27–30
high resolution and, 4.86–87
image, 10.232–233
mirror electron microscopy, 4.167–207
photographic, 1.187, 190
polarizing microscopy, 1.83
scanning electron microscopy,
  83.205–206, 208–209, 213, 219,
  234–235
 backscattered electrons, 83.215,
   225–226
 beam voltage, 83.209, 235–236
 cathodeluminescent, 83.205, 213
 Monte Carlo, electron scattering
   simulations, 83.218–219
 topography, 83.206, 213, 218, 231,
   236
 types, 83.205–206
simulations, electrical microfield,
  94.135–140
transfer, 4.75–79
Contrast detail curve, 84.340
Contrast embedding, 6.228
Contrast formation, basics, 87.133–138
Contrast-stop imaging, 2.206
Contrast stretching, 92.4–6
Contrast transfer function, 86.175
Control parallelism, 87.269
CONV (software), 13.85, 91
Convergence:
 power series, 90.82
 regularity-free convergence of multigrid
   methods, 82.343
Convergence bound, 90.81
Convergent-beam electron diffraction
 (CBED), 11.47; 86.175, 197;
 90.211, 214, 218–221
 apparatus, 96.588–589
 Gottfried's research, 96.585–586,
   588–591, 593–595
 imaging plate and, 99.269–270
 intensity minima, 96.594–595
 Japanese contributions, 96.766
 Kikuchi lines, 96.585, 593–596
 Kossel effect, 96.585

specimen preparation, 96.591, 593
Convex feasibility problem; *See also*
  Convex set theoretic image
  recovery
 convex feasibility in a product space,
   95.171–172
 convex functionals, 95.165
 Fejér-monotone sequences, 95.170–171
 geometric properties of Hilbert space
   sets, 95.162–163
 inconsistent problem solution:
  alternating projections in a product
   space, 95.203–206
  least-squares solutions, 95.202–203
  simultaneous projection methods,
   95.206–209
 nonlinear operators, 95.168–170
 projections:
  distance to a set, 95.166
  operators, 95.166–167
  relaxed convex projections, 95.168
  steps in solution, 95.161, 199–200
  topologies of subsets, 95.163–164
Convexity, 91.46–48, 62–70
 majorization, 91.47
 maxpolynomials, 90.107
 in pairs, 91.68–70
 pseudoconvexity, 91.47, 66–67
 quasiconvexity, 91.47, 67
 Schur-convexity, 91.48, 67–68, 70
Convex set theoretic image recovery:
 affine constraints, 95.179, 259
 applications, 95.180–181, 183–184
  restoration problems, 95.182
  tomographic reconstruction,
   95.182–183
 basic assumptions, 95.172
 confidence level, 95.198–199
 history, 95.176–180
 image space models:
  analog, 95.173
  digital, 95.173–174
  discrete, 95.173
  general, 95.172–173
 information management, 95.198–199

Convex set theoretic image recovery, *continued*
  nonconvex problem solving
    convexification, 95.184–185
    feasibility with nonconvex sets, 95.186
    new solution space, 95.185
    property sets in Hilbert space, construction, 95.189–198
    set theoretic formulation, 95.174–176
Convex structuring functions, 99.13–14
Convolution, 88.301, 311; 93.144
  by sections, 94.4–8
  cyclic convolution, 94.2, 4–7, 19
  fast convolution, 94.12–19
  filtering, 94.2–3, 4–8
  group convolution, 94.3, 11–12
  linear convolution, 94.2, 4–7, 12, 19
  vector convolution, 94.11
  Walsh convolution, 94.19
Convolution kernel, 92.55–58
Convolution paths, 90.373–379
Convolution property, matrices, 94.29–35
Convolution square pattern, 93.223
Convolution theorem, for Fourier transforms, 86.190, 191, 210
Cook Tom algorithm, 94.18
Cooley Tukey (CT) algorithm, 93.17–19, 27–30
  abelian affine group, 93.49–50
  abelian point group, 93.47–49
  affine groups, 93.3, 46–47
  extended, 93.47–52
  multidimensional, 93.28–30
Cooling chamber, 1.161, 162, 170
Cooperative integration, 86.86, 125, 139, 142
Coplanar waveguide, 83.78
Copper:
  crack development in stressed copper plate, 98.21, 22–30, 31–33
  melting point, 98.6–9
Copper/cobalt, magnetic multilayers, 98.406–408
Copper foil, crystal-aperture scanning transmission electron microscopy, 93.59, 90–91
Copper oxide, 91.142
Copper phthalocyanines, 5.314, 316
Core-loss peaks, 11.89
Cornu's spiral, 13.260
Correctionism, 94.265, 292
Correlated noise, 85.24, 66, 68
Correlated uncertainty process, 95.197–198
Correlation coefficient, 85.15, 38
Correlation energy, 82.199, 210
Correlation function, atomic displacements, 90.286
Correlation quality, 7.139, 140
Correlator, multiplexed, optical symbolic substitution, 89.82–84
Corrosion, wet replication study of, 8.99
Corynebacterium diphtheriae, 5.335
Cosmological principle, information approach, 90.170, 188
Cossor Electronics, 91.234, 238–239
Cost computation rule, LLVE, 86.121, 123
Coulomb gauge, 82.2, 18, 20, 26, 28, 30, 34, 35, 41, 42, 43
Coulomb interaction, 13.216–231
  Boersch effect, 13.216–220
  closest encounter techniques, 13.218
  and electrostatic charging, 13.224–231
  reducing effects of, 13.222–224
  reducing trajectory displacement effect, 13.222–224
  trajectory displacement effect, 13.220–222
Coulomb problem:
  Hamiltonian, 95.306
  spacetime algebra solution, 95.306–307
Coulomb scattering, spacetime algebra, 95.313–314
Coupled slot, surface field, 92.110–111
Coupling:
  of eddy current and static magnetic field, 82.29
  electromagnetic:
    coupling coefficient, 89.213–214

optical interconnects, 89.210–213
quantum devices, 89.209–243
quantum mechanical:
  between quantum wells, 89.195–199
  coupling coefficient, 89.213–217
  shortcomings, 89.221–223, 244–245
  spin-phonon, 89.241–224
Coupling coefficients, 89.213–217
Covariance, 85.15, 19, 64, 68
  information approach, 90.186
  inverse Toeplitz covariances, 84.271–272
  models, 84.264–268
  quantum mechanics, 90.154–156
Covariance estimation, abstract algebra and, 84.262–313
Covariance matrix estimate, 85.20–25, 33–48, 50–55, 63
  explicit, 84.282, 284, 287
  general sample, 84.264–265, 268, 292–293
  inverse linear, 84.301
  isomorphic block diagonal, 84.280
  linear, 84.263, 269, 271, 282
  normal equations, 84.262, 289, 291, 297
  orthogonal subspace decomposition, 84.278
Covariance method, speech coding, 82.124
Covariant equation, 85.3–4, 36–37, 40
Covariant graph, 85.56, 65
Covering, group theory, 93.10, 16
Cowpea mosaic virus:
  chlorotic, 6.263, 264, 266
  reconstruction, 7.370, 372, 373
CPU time, 85.29, 62–64, 70
Crack code, of a contour, 85.103
Cracks:
  image analysis, 2.220; 84.200, 203, 218
  quantitative particle modeling, 98.21, 22–30, 31–33
Cramèr–Rao inequality, 84.286–287, 292, 302, 304–306, 310; 90.129–133, 165–166, 177–178, 187; 94.366, 371–372, 376
Crank–Link, 3.21

Cray computers, 85.271–297; *See also* Parallel programming
  architecture, 85.271–273
  autotasking, 85.292–297
    code generation, 85.297
    data dependence analysis, 85.292–294
    translation, 85.294–297
  languages, 85.273–275
  macrotasking, 85.275–286
    events, 85.279–286
    locks, 85.278–279
    task declaration, 85.276–277
  microtasking, 85.286–292
    compiler directives, 85.286–287
    rules for using, 85.290
  operating systems, 85.273–275
  operating systems and languages, 85.273–275
Critical concentration $NSUBcsub$, 82.205, 206
Critical current density in high-temperature superconductivity, 14.31–38
Critical cycle, minimax algebra, 90.41
Critical diagram, minimax algebra, 90.22
Critical events, minimax algebra, 90.16–26
Critical field, in high-temperature superconductivity, 14.31–36
Critical illumination, 7.141; 10.7
Critical path analysis, 90.2–23
Critical-point drying method, Japanese contributions, 96.783
Critical temperature, 83.13
Cross-correlation, 84.143–144
  generalized, 84.151
  Mellin-type, 84.158
  normalized, 84.143
  visual perception, 84.142–143
  wave function, 7.108
Crossed fields, parasitic effects, 86.239, 278, 279
Cross-grating Fourier transforms, 8.16
Crossover:
  electron gun ray pencils, 8.140–144, 172–174
  field electron emissions, 8.245, 251

Cross-spectrally pure fields, 7.110
CRT *See* Cathode ray tube
Cryoelectron microscopy, Japanese contributions, 96.739
Cryo-lenses, 10.239–241
Cryo-techniques, 83.252–255
Crystal-aperture scanning transmission electron microscopy (STEM), 93.57–107
  direct imaging of nucleus, 93.87–89, 104
  experimental, 93.66–87, 90–91
  imaging, 93.58–59, 63–66, 87–90, 94–106
  resolution, 93.91–94
  theory, 93.59–66
Crystal disorder:
  binary solids:
    lipids, 88.181, 183
    paraffins, 88.180
  effect on diffraction intensities, 88.174
  phase transitions:
    cholesteryl esters, 88.177
    fluoroalkanes, 88.178
    paraffins, 88.176
    porin, 88.178
Crystal growth, 2.401
  epitaxial orientation:
    on inorganic substrates, 88.119
    on organic substrates, 88.120
  Japanese contributions, 96.603–604, 608–610, 698–700, 751–752
  Langmuir Blodgett films, 88.118
  protein reconstitution, 88.120
  solution crystallization:
    evaporation of solvent, 88.118
    self-seeding, 88.118
Crystal lattice:
  electrons, transmission through, 93.59–62, 66–73
  zone axis tunnels, 93.73–79
Crystalline axes, 1.92
Crystalline fluorescence, 1.73
Crystallographic phase determination:
  direct methods:
    crystal bending, effect of, 88.153
    density modification, 88.141
    dynamical scattering, effect of, 88.149
    electron microscope images, 88.130
    examples of solved structures, 88.139
    maximum entropy, 88.142, 171
    phase invariant sums, 88.134
    Sayre equation, 88.141
    secondary scattering, effect of, 88.151
  Patterson function, 88.132
  trial and error, 88.131
Crystallographic residual, 88.131
Crystallography:
  group-invariant transform algorithms, 93.1–55
  X-ray, 7.189, 245; 93.167–168
Crystal refractometer, 1.75
Crystals; *See also* Crystallography
  bending:
    diffraction incoherence from, 88.152
    effect on structure analysis, 88.153
  color schlieren system photographs, 8.14
  deformed, 90.238–241
  disorder *See* Crystal disorder
  dynamical elastic diffraction, 90.221–250
    boundary conditions, 90.226–230
    equations, 90.221–224
    reflection amplitude, 90.229–230
    reflection high-energy diffraction, 90.241–250
    transmission high-energy electron diffraction, 90.236–241
    two-beam approximation, 90.230–236
  elastic scattering, 90.216
  electron microscopy, 7.288
  electron off-axis holography, 89.37–38, 44–47
  geometry, 1.42
  growth *See* Crystal growth
  GS algorithm and, 7.222
  inelastic scattering and, 7.264
  lattice *See* Crystal lattice
  lattice spacings, 4.81
  mirror electron microscopy, 94.96–97
  morphological directions, 1.70

mounting of, 1.46, 54, 55
multilayer system, 90.238
optical properties, 1.41
phase information and, 273
protein molecules in, 7.330–332
selvage, 90.241, 245–248
semi-infinite, 90.235, 243–248
slab, 90.229–230
RHEED, 90.248–250
structure *See* Crystal structure
substrate, 90.241, 243–245
Crystal spectrometer, 6.281
Crystal structure:
of chitosan, 88.137, 138, 165
of copper perchlorophthalocyanine, 88.161
determination, 88.115; 90.265
linear least-squares method, 90.266–267
nonlinear least-squares method, 90.268–272
of diketopiperazine, 88.157
direct inversion, 90.257–265
linear model, 90.259
quadratic model, 90.260–265
direct methods, 88.139
factors, 90.257–265
luminescence and, 99.243
membrane proteins:
bacteriorhodopsin, 88.168
porins, 88.168–169
methylene subcells, 88.155
of paraffins, 88.162
of phospholipids, 88.163
of poly(butene-1), 88.167
poly(E-caprolactone), 88.166
of polyethylene, 88.166
polymer structures, 88.133
potential and, 90.338–341
CSF electron microscopes, 10.225, 228; 96.430, 437, 446, 504
Csiszár's phi-divergence, 91.41, 116, 117; 97.120
CT algorithm *See* Cooley Tukey (CT) algorithm

CUBE system, 85.163–165, 176
Cumulants, 94.323, 324
Current amplification, 83.21
Current density:
displacement, 82.42
electromagnetism, 82.2, 8, 12, 19, 27, 29, 77, 85
electron field emission, 95.74–76, 78
electron gun, 8.200
in Schrödinger equation, 13.254–255
space charge limited, 83.49
surface:
electric, 82.8
magnetic, 82.5, 11
vacuum microelectronics, 83.5, 7, 10, 21, 28, 47, 57, 75–77
Current noise, 87.206
Current vector potential, 82.19, 27
Cuts, fuzzy set theory, 89.264–265
Cycle means, 90.40–41
Cyclic convolution, 94.2, 4–7, 12–19
Cyclic groups:
index notation, 94.8–9
matrix representation, 94.13
Cyclotron:
effective mass, 92.81
frequency, 92.81
mobility, 92.87
radius, 12.110
resonance, 92.89
CYDAC, 2.79, 81, 145; 5.43, 44, 45
electronic design of, 2.89, 91, 93, 111
performance, 2.96
specification of, 2.82, 83
Cylinder, computer modeling, 13.29–30, 34
Cylindrical electrostatic analyzer, 4.267–287, 319–324, 340–342
Cylindrical magnetic analyzer, 4.287–302, 324–327
Cylindrical mirror, 86.246, 279
Cytoanalyzer, 4.362
Cytochrome oxidase, 6.212, 216; 8.123
Cytochrome reductase, negative staining, 8.123

Cytological smear preparations, stripe
    photomicrography, 7.81
Cytologic specimens, contour lines,
    8.20
Cytoplasm, 6.149, 181, 182, 230
Cytoplasmic granules, 6.77

D

DAB technique *See* Diaminobenzidine
    technique
Dacite, microtransform, 7.48
Damage, radiation *See* Radiation damage
Dark currents, 2.10; 99.74
  avalanche photodiodes, 99.88, 150–153,
    156
Dark-field, 11.25, 154
Dark-field/bright-field illumination, 6.21;
    7.273
Dark-field electron microscopy, 6.116
  of biological specimens, 5.299,
    300–303, 342
  disadvantages, 5.303
  for image improvement, 5.102, 299,
    300–303
  phase determination and, 7.274
  phase problem and, 7.243
  resolution, 5.342, 343
  signal-to-noise ratio, 7.257
  strioscopy, 5.300
Dark-field illumination, 6.17, 18, 21
Dark-field images, 2.206, 212, 232; 10.25
  formation, 8.123
  phase problem solution and, 7.227,
    228
Darwin's solution, electron diffraction
    theory, 90.236
Data, granularity, rough set theory,
    94.151–194
Database, fuzzy relations, 89.313–314
Data compression, 94.386–391
Data formation model:
  image recovery problem, 95.156–157
  set construction, 95.190–191

Data parallelism, 87.269
Data processing, 2.77, 119
Data structure, 88.63
  design in C, 88.73
  implementation, 88.81
    BOUNDARY structure, 88.85
    DEVICE structure, 88.89
    IMAGE structure, 88.82
    K<->STATE structure, 88.88
    POLYLINE structure, 88.89
    scalar types, 88.90
    SEQUENCE structure, 88.86
    TRACE structure, 88.87
    VIDEODISC structure, 88.87
    WINDOW structure, 88.84
  pyramid, 86.133, 138
  spatial, 86.147, 164
Daugman's neural network, image
    reconstruction, 97.31, 52
d-band emission, 8.218
DCT *See* Discrete cosine transform
DD *See* Direct detection
Deadtime, relation to counting rate, 6.285,
    286
Deblurring
  image reconstruction, 97.155–159
  space-variant image restoration,
    99.315–318
De Broglie, Louis, 12.17, 18
Debye–Hückel approximation, 87.108
Debye length, 82.225; 87.107
Decimation, group theory, 93.8, 15, 16
Decision tables:
  Dempster rule of combination, 94.176
  rough sets, 94.154–157, 158, 185–187
Decision tree, 84.238, 241, 243
Declarative analysis, 86.81–86, 89, 91, 93,
    144, 146, 154, 163–164, 166, 188
Decoder, digital coding, 97.192
Decomposition models, 85.96
Deconvolution, 87.5–6
  blind, 93.144–152
  fully-connected network and, 87.11 ff.
Deep level transient spectroscopy (DLTS),
    10.67

Defects:
  crystals, electron off-axis holography, 89.44–47
  diffuse scattering, 90.282–285
  electrical noise, 87.216, 219, 222, 224–225, 232
  mirror electron microscopy, 94.97
  photomicrography, 7.67
  point, 87.219, 225
Definable partition, 94.167–171
Definable set, 94.158–159, 166
Definite matrix, 90.56–58
Deflection, 83.152
  aberration, 83.156–162
  cardinal elements of, 13.187–188
  first order, in electron trajectories, 13.186–188
  fundamental trajectories, 13.187
  inhomogeneous stratified media in, 8.9
  post-lens single deflection, 83.152–156
  pre-lens double deflection, 83.152
  pre-lens single deflection, 83.162
  refractive index gradient relationship with, 8.8
  system for SEMs, 13.113–119
  yokes, location, 83.152–156
Deflection bridge, mirror EM, 4.212–214
Deflection fields:
  in electron trajectories, 13.186–188
  electrostatic, 13.146–149
  magnetic, 13.149–152
  in SEMs, 13.146–152
Defocus error, phase problems and, 7.262
Defocusing, 5.163, 165, 174, 179, 196–198, 240, 241, 245, 249, 294; 86.233, 239, 243, 266, 269
  elastic image and, 7.264
  of electron image, 1.172
  in electron optical transfer function, 13.274–277
  and axial astigmatism, 13.278–279
  and rotationally symmetric third-order aberration, 13.280–284
  thin carbon film image and, 7.186
Deformed crystals, 90.238–241

Degenerate Fermi–Dirac system, 82.208
Degree of coherence, measurement, 7.158–163
Degree of local spatial order, 92.45
Degree of polarization of fluorescence:
  defined, 8.55–56
  sign, 8.56
  significance, 8.56–58
Degree of spatial coherence *See* Complex degree of coherence
Degree of uncertainty, rough sets, 94.181
Dehydrogenase, 6.209
  electron microscopy, 6.209, 214
  ferricyanide technique, 6.210
  histochemistry, 6.210
  tetrazolium salt technique, 6.209, 210, 212, 213, 214
Delaunay triangulation, 86.147
Delay, minimax algebra, 90.6–7
Delayed-choice experiment, 94.297–298
Delesse principle, 3.69
Delft particle optical system, 13.154, 217
Delocalization, 83.240, 241
Δ, minimax algebra, 90.54–58
Dember voltage, 10.69
Demodulation, 84.328
De Morgan's law, 89.231, 259
Dempster rule of combination, 94.176
Dempster–Shafer theory *See* Evidence theory
Denisyuk hologram, 10.139
Denoising, image enhancement, 97.56–58
Dense electron gas, 82.199, 207
Density:
  of image, relation to exposure, 1.183
  microkymography and, 7.92
  SEM and, 83.206, 209, 218, 225, 231, 249, 251
  z contrast image, 83.215–216, 83.225, 227, 233, 247–248, 251
Density of states, 82.204, 213, 227, 232, 237
  in band tails, 82.201
  deep tail, 82.242
  distortion of, 82.201, 245

Density of states, *continued*
   effective mass, 82.203, 204, 209, 222
   effect of, 82.198
   in heavily doped silicon, 82.199, 200
   in the tails, 82.240
Deposition, 83.14, 54
Depth cue, 85.85, 114
Depth of focus, 3.27; 83.110
Depth profiling, 11.147
Derivative spectroscopy, 8.44
Derjaguin approximation, 87.67
   electric force microscopy, 87.130
   geometrical errors, 87.93
   geometry, 87.65–66
Derjaguin–Landau–Verwey–Overbeck theory, 87.110
DES *See* Discrete-event system
Descriptors, 3D display, 85.97, 99, 101, 103, 105
Detective quantum efficiency (DQE), 12.57; 89.33, 34; 99.260–261
Detectors, 83.205
   backscattered electron, 83.206, 223–226
   bandwidth, 83.223, 225–226
   dead layer, 83.225
   efficiency, 12.100
   electron-counting, 83.257
   improvement, 83.223
   low-beam voltage, 83.223
   microchannel plate (MCP), BSE, 83.225–226
   misalignment caused by, 83.223–224, 230
   noise, digital image processing, 4.141–145
   position sensitive, 83.225
   secondary electron:
      Everhart–Thornley, 83.222–223, 230
      TEM/SEM, 83.222
   semiconductor, 83.225
   sensitivity, tracking microscope, 7.12
   STEM, geometry, 7.143
DFT, 88.6
DIA-Expert, 86.100
Diagonal realization, DES, 90.108–109

Diamagnetic flux shield lens:
   calculation of configuration, 5.229, 230, 231
   characteristics, 5.229
   construction, 5.228, 232
   electron microscopy, 14.20–21
   flux density, 5.229, 230, 231
   flux density prediction, 5.232
   flux jumps, 5.233
   misaligned, 5.233
Diaminobenzidine technique, 6.210–212, 220
Diamond, 5.308
Diamond powder, 5.153, 154, 155
Diatomic molecules:
   KFS phase contrast microscopy, 6.80
   melting points, 98.13, 14
   molecular bonds, 98.67–74
DIC *See* Differential interference contrast
Dichroic ratio, 8.65–66, 66–72
Dichroism, 1.72, 88, 93
Dictyostelium discoideum, 5.336
Dielectric constant, 82.3–8, 10–11, 15, 27, 41
Dielectric function, 82.210, 230
Dielectric materials:
   multiple cuts in FDM methods, 13.21–23
   single cuts in FDM methods, 13.11–21
Dielectric particles, 83.12
Dielectric permittivity, 87.56
   as function of frequency of water, 87.73–74
   solvation forces, 87.115
Dielectric spheres, charged, electron holograms, 99.174, 207–216
Diffeomorphism, 84.181
Difference equations, 13.7–8
   for dielectric materials, 13.12–21
   for magnetic materials, 13.24–25
   for multiple dielectric cuts, 13.21–23
   in ray tracing, 13.206
   solving, in FDM methods, 13.10
Difference of Gaussian (DOG), receptive field, 97.44–45

Differential algebraic method, 91.3
Differential amplitude imaging, 10.54
Differential blood counting, 9.348–350
Differential capacitive position sensor, thin film stack pattern, 14.82–83
Differential cell refractometer, 1.57, 59, 70
Differential cross-section, scattering experiments, 90.208
Differential equation:
  of current vector potential, 82.19, 20, 27, 28
  eddy current field, 82.21
  electric scalar potential, 82.10, 26, 27, 41, 42
  electric vector potential, 82.42, 43
  elliptic, 82.43
  Laplace, 82.17, 18, 34
    generalized, 82.10, 12
    vector, generalized, 82.18
  magnetic scalar potential, 82.12, 14, 27, 28, 42, 43
  magnetic vector potential, 82.16, 17, 26, 27, 41, 42
  parabolic, 82.43
  partial, 82.2, 43
  set of, ordinary, 82.47, 48
  static field, 82.4, 5, 7, 8
  waveguide and cavity, 82.39, 45
Differential equation of states (DES), 94.336
Differential interference contrast (DIC), 6.105
Differential operators, 82.44, 45; 94.335
Differential phase contrast (DPC) mode, 10.49–55
  geometrical optics, 98.351–357
  imaging, 10.52
  wave optics, 98.357–358
Differential power law index, 87.70
Differential pulse code modulation (DPCM), 82.153; 97.51
Diffraction, 11.28; 83.207, 225; 84.319, 332
  anomalous diffraction, 11.35
  Bragg diffraction, 84.322
  charged-particle beam:
    scalar theory, 97.279–282
    spinor theory, 97.330–332
  correction, 84.323
  crystal-aperture scanning transmission electron microscopy, 93.63–65
  double diffraction, 11.82
  dynamical elastic diffraction *See* Dynamical elastic diffraction
  effect, 84.344
  electron diffraction *See* Electron diffraction
  Fraunhofer diffraction *See* Fraunhofer diffraction
  Fresnel diffraction *See* Fresnel diffraction
  in microscope objective, 14.254–255
  term, 84.323
  theory, 5.260, 261
  X-ray, 4.113, 115, 118
Diffraction contrast, Japanese contributions, 96.601–602
Diffraction electron microscope, 91.278–279
Diffraction grating, 1.28, 29
  as holograms:
    amplitude, 10.117
    classification, 10.139–140
    diffraction efficiency, 10.121
    Lipmann–Bragg, 10.126
    phase gratings, 10.122, 124
    reflected light, 10.122, 125
    "sinusoidal" profile, circular zone plate, 10.118–123, 129
  optical symbolic substitution, 89.72–73
Diffraction–image plane method, phase determination and, 7.273
Diffraction integrals:
  calculation of:
    for spherical aberration, 13.261–264
    for spherical aberration-free condition, 13.265–268
  stationary phase approximation, 5.263, 264, 272
Diffused transistor, 91.145

Diffuse scattering, 90.217–218, 279–285
  defects diffuse scattering, 90.282–285
  real diffuse scattering, 90.348–349
  virtual diffuse scattering, 90.348–349
Diffusion, 83.16
Diffusion current, 86.13
Diffusive transport, 89.107
  Aharonov–Bohm effect-based devices, 89.151–152
Digital coding, 97.192–194
Digital differential analyzer, 85.127–128
Digital filtering, 85.16
Digital frame store, 11.155
Digital image processing, data recording, 4.149–153
  conversion to output, 4.150–153
  display, 4.153
  electronic recording, 4.149–150
Digital images, 85.87; *See also* Image enhancement; Image processing; Image reconstruction
  acquisition, 93.229
  edge, 93.231
  noisy, 93.257, 277
Digital integrator, 91.152
Digital optical receiver, 99.72
Digital revolution, telecommunications, 91.195–197
Digital storage, 10.9
Digital straight segments (DSS), 84.225–226
Digital system, 91.195; *See also* Image enhancement; Image processing; Image reconstruction
  electron microscopy, 89.6
  image processing, 4.127–159
  spin-polarized single-electron device, 89.233–234
Digital tapes, conversion to visual presentation, 4.89
Digitization, 84.250, 254
Dihedral groups:
  index notation, 94.9
  matrix representation, 94.13
  matrix transforms, 94.32–35

Dilation:
  function, 88.203; 99.8
  mathematical morphology, 88.202, 203, 231, 329; 89.328–329, 343–344
  multiscale dilation–erosion scale-space, 99.16–22
Dimension, of space elements, 84.202
Dimensional funtionals, 99.41
Dimensionality, 85.91; 90.163; 99.8, 40–42
Diode array detectors, 9.143, 144, 326, 329
Diode lens, electron optical aberrations, 8.257
Diodes *See* Avalanche photodiodes; Field emission diodes; Light-emitting diode; Photodiodes; Tunnel diode; Zener diode
Dirac equation, 90.154, 161–164, 179, 187
  CPT symmetry handling, 95.295
  Dirac adjoint, 95.292–293
  Foldy–Wouthuysen representation, 97.267–269, 322, 341–347
  Hamiltonian form, 95.299
    angular momentum operators, 95.302
    nonrelativistic reduction, 95.299–301
    notation, 95.298
    Pauli theory, 95.299–301
    spherical monogenics, 95.303–305
  Hermitian adjoint, 95.292–293
  observables, 95.292–296
  plane-wave states, 95.295–296, 316
  spacetime algebra version, 95.292, 297
  spinors, 95.293–296
Dirac matrix, 84.277
Dirac oscillator:
  equation, 95.307
  Hamiltonian, 95.307
  negative kappa equation, 95.308–309
  positive kappa equation, 95.308
Dirac plane wave:
  evanescent waves, 95.316, 320–321
  Klein paradox, 95.328–330, 332
  matching at a potential step:
    evanescent waves, 95.321–322
    oblique incidence, 95.315
    perpendicular incidence, 95.315

traveling waves, 95.318–320
spacetime algebra equations,
    95.295–296, 316
spin precession at a barrier, 95.323
    polarization operator, 95.324–325
    precession angle, 95.323–324
    reflection coefficient, 95.323
    rest-spin vector for reflected wave,
        95.323
    traveling waves, 95.316–317, 318–320
    tunneling time *See* Tunneling
Dirac spinor:
    complex conjugation, 95.291–292
    Dirac–Pauli matrix representation,
        95.289–290
    operator action of matrices, 95.288–289
    vector derivative operator, 95.280–281
    Weyl matrix representation, 95.290–291
Direct detection (DD), 99.68
Directed graphs, 90.36–41
Direct electron ray tracing, software for,
    13.39–41
Direct inversion, crystal structure factors,
    90.257–265
Directional couplers, 89.193–199
Direct product group algebra, 94.14–19
Direct reconstruction, three-dimensional
    images, 7.304, 314–324
Direct tensor product, group algebra,
    94.16
Dirichlet boundaries, 82.11, 12, 14, 18, 27,
    44, 46, 51, 52, 54, 58, 62, 63, 67,
    70, 85
    in FDM methods, 13.34–35, 36
Dirichlet problem, 8.226, 237, 251; 87.140
Disc of least confusion, 11.135
Discontinuities:
    image processing, 97.89–91
    image reconstruction, 97.108–118
        duality theorem, 97.91, 115–118
        explicit lines, 97.110–115, 154–166
        implicit lines, 97.108–110, 166–181
        line continuation constraint,
            97.130–141, 142
Discontinuous potential, 86.251

Discrete cosine transform (DCT), 88.2, 7,
    10; 97.11, 51, 52, 97, 194
    four versions, 88.12
    integer *See* Integer cosine transform
    weighted, 88.9
Discrete-event system, 90.2–4
    connectivity, 90.45–58
    critical events, 90.16–26
    efficient rational algebra, 90.99–119
    infinite processes, 90.75–84
    maxpolynomials, 90.88–98
    orbit, 90.13–14
    path problems, 90.36–45
    Period-1 DES, 90.110–111
    realizability, 90.117–119
    scheduling, 90.26–36
    steady state, 90.58–74
    strong realization problem, 90.108–109
Discrete linear systems, signal description,
    94.322
Discrete matrix transform, group algebra,
    94.27–44
Discrete representation, 85.98
Discrete signals, exact Gabor expansion,
    97.30–33
Discrete sine transform (DST), 88.8, 28
    the four versions, 88.12
    integer, 88.27
Discrete spectrogram, 97.11
Discrete transistor, 91.146
Discretization:
    reciprocal space, 7.213
    three-dimensional reconstruction, 7.305,
        307
Discretization noise, distance transform to
    smooth, 92.72–75
Disjunction, fuzzy sets, 89.261, 297
Dislocations, 2.233, 382, 384, 400; 87.222,
    231, 233
Disorder:
    Fisher information as measure,
        90.136–137
    information approach, 90.175, 186
Dispersion, 86.239, 243, 266
    characteristics, 82.86, 87, 90

Dispersion, *continued*
  mirror-bank energy analyzers, 89.397, 403
  of optical directions, 1.72
  of refractive indexes, 1.70
Dispersion force, nonretarded, 87.76–77
Dispersion relations, chiral media, 92.125–129
Dispersion shifted fiber (DSF), 99.71
Displacement current, 83.48
Displacement meter, 91.175
Displays, 83.6, 78, 80–82
Disputer array, 87.287
Dissipant, 94.328, 329–330
Distance measure, 87.36 ff.
Distance transform, binary image enhancement, 92.72–75
Distorted wave approximation (DWA), 90.273–275
Distorted wave Born approximation (DWBA), 90.217, 218, 275, 279
Distortion, 2.5; 14.267; 82.137
Distortion fields, mirror electron microscopy, 94.109–124
Distributed problem solving, 86.139, 142
Distribution:
  autoregressive process parameters, 84.297–300
  convergence, 84.309
  exponential family, 84.263, 274
  Gauss–Markov, 84.271
  heavy-tailed, 84.309
  multivariate Gaussian, 84.264, 274, 380
  sample set, 84.264, 297, 299
  univariate Gaussian, 84.297
  Wishart, 84.280–281
Distribution artifacts, 8.114–115
Dither, 85.6
Dithered-beam metrology, 8.44
Divergence:
  vector potential, 82.17, 35, 36
    normal derivative of, 82.18, 27, 28, 30, 37
Divergence measures, 91.37–38, 41
  A- and G-divergence, 91.40
    M-dimensional generalization, 91.79–82
  arithmetic-geometric mean divergence, 91.38
  I-divergence, 91.40
  J-divergence, 91.37, 113
    Bhattacharyya's distance, 91.131–132
    M-dimensional generalization, 91.40, 71–79
  Jensen difference divergence, 91.38
    M-dimensional case, 91.38–40, 76–82
  T-divergence, 91.40
    M-dimensional generalization, 91.79–82
Diverging lens, 11.106
Division algorithm, 84.115; *See also* Skeletonizing technique
DLP theory, 87.54–56, 59, 82
  macroscopic, 87.61–62
DLTS *See* Deep level transient spectroscopy
DNA, 5.302, 303, 329
  anistropy, 8.67
  arrangement interpretation, 8.66–72
  axial ratio, 8.51
  degree of polarization of fluorescence, 8.66–72
  dehydration, 5.329
  dichroic ratio, 8.65–66
  dipole orientation field distribution, 8.70
  double-stranded, contour length measurement, 8.127
  dye binding, 8.53, 66–67
  electron microscopy, 7.271; 8.126–127
  first-order coil, 8.68, 69
  hybridization of, fluorescent probes for, 14.172
  hydrophobic grooves, 8.67
  interaction with solid surfaces, 8.117–119
  intercalation, 8.53, 56
  ionic charge, 8.109–110, 112
  linear dichroism, 8.63, 64, 66–72
  molecule length, sample preparation method effect on, 8.119

nucleosomes determination, 8.126
optical anisotropy, 8.52, 56
orientation, 8.64, 66, 70
packaging direction, 8.51–52, 67
photoelectron microscopy, 10.311–312
radiation damage, 5.318
single-stranded, spreading of, 8.127
spreading adsorption on various
  surfaces, 8.117–119
stacking orientation, 8.64
structure, 7.288
supercoiling, 8.68
  equations for, 8.68–69, 70
  objective lens numerical aperture
    effect, and, 8.69–70
  photograph, 8.71
ultraviolet absorption, 8.63
uranyl stained, images, 7.234, 235
zero-order, 8.67, 68
DNA–acridine orange complex, 8.56–60
DNA–dye complexes
  depolarization, 8.63
  DNA packing materials, 8.59
  dye-binding characteristics, 8.59
  orientation, 8.54, 56, 60–63
  partial polarization of fluorescence,
    8.60, 63
  polarized fluorescence microscopy,
    8.53–63
  preparation, 8.58–60
  quantum yield, 8.54
  relaxation time, 8.62
DOD CELP, 82.175
DOD standard, speech coding, 82.175
Domain, 93.227, 287
  decomposition, 82.362
  reduction, 82.356
Domain segmentation, 93.287–299, 325
  algorithms, 93.304–307
  anisotropy index, 93.298, 299
  image presentation, 93.296–298
  modal filter, 93.288–292
  multispectral analysis, 93.318–319
  radius, choice, 93.293–296
  Rayleigh statistical test, 93.292–293

vector magnitude, 93.293
Domain structure:
  magnetic materials, 3.183–190; 98.324,
    325–327, 388–389
  mirror electron microscopy, 94.97
Domain walls, 98.327–328
  converging image, 5.275, 276, 282, 293,
    294
  diverging image, 5.274, 275, 276, 277,
    292
  Fresnel images, 5.282, 283
  interference fringes, 5.254, 255, 293
  measurement, 5.240, 291–294
Donors, vacuum microelectronics, 83.8, 9,
  20; 87.225
Doping, 91.144, 185–186
Double-barrier resonant tunneling device,
  89.130–135
Double channel analyzer, 12.174, 177
Double diaphragm method, 1.65, 66, 67,
  73, 74
Double diffraction, 11.82
Double-exposure electron holography,
  99.205–207, 207–210
Double layer, electric, 82.5
Double orbit table, 90.22
Double-ordered cavity, three-dimensional
  microstrip circulators, 98.283–287
Double-passage coherent imaging,
  93.152–154
Double quantum wells, Aharonov–Bohm
  effect, 89.149–178
Double quantum wire Aharonov–Bohm
  interferometer, 89.172–178, 180,
  186, 199
Double sideband holography, 7.251
Dove prism, 1.15, 38
DPCM *See* Differential pulse code
  modulation
DQE *See* Detective quantum efficiency
Dreams, script, 94.290
Drift:
  high voltage supply, 1.125
  thermal, 1.150
Drift diffusion formalism, 89.110–113

Drift error, 85.7
Dropping conditions, 94.188
Drops:
   collisions of microdrops, 98.13–21, 22
   formation on solid surfaces, 98.30, 32–44
   motion within fluids, 98.44–61
Drosophila melanogaster chromosome, 6.167
Drude model, 92.114
Drying forces, structural damage produced by, 8.116–117
Dry joint, 87.234, 240
Dry mass determination, 6.135, 136
Dry organic matter determination, 1.99
DSF *See* Dispersion shifted fiber
DST *See* Discrete sine transform
Dual basis, 89.348–349, 354–358
   gray-scale τ-mappings, 89.371–372
   translation-invariant set mappings, 89.363
Dual conorm, fuzzy set theory, 89.261
Dual covering, group theory, 93.10
Dual gray-scale reconstruction, 99.49
Dualist theory, mind-body problem, 94.261–262, 294, 305–306
Duality, abelian group, 93.3, 7, 8
Duality theorem:
   image processing, 97.91, 115–118
   minimax algebra, 90.26
Dual rail coding, 89.68–69
Dual transportation problem, image algebra, 84.124
Duker, Heiner, 12.10
Duoplasmatron, 11.106
DWA *See* Distorted wave approximation
DWBA *See* Distorted wave Born approximation
Dwell time, 89.136
Dyadic recursive Green's function, 92.103–106, 108, 112
   microstrip circulators:
      three-dimensional model, 98.219–238
      two-dimensional model, 98.79–81, 98–108, 121–127, 316

Dyadic symmetry:
   decomposition, 88.23–25
   defined, 88.17
   dependence, 88.20
   type, 88.18, 20
Dye binding, 8.53–54
   DNA arrangement, effect on, 8.66–67
Dye compounds:
   metachromasy of fluorescence, 8.53
   purification, 8.58
Dynamical elastic diffraction, 90.221–250; 93.58, 60
   boundary conditions, 90.226–230
   equations, 90.221–224
   reflection amplitude, 90.229–230
   reflection high-energy diffraction, 90.241–250
   transmission high-energy electron diffraction, 90.236–241
   two-beam approximation, 90.230–236
Dynamical phase effects, 89.38–44
Dynamical scattering, 88.147–149
   multiple beam theory, 88.148
      phase grating approximation, 88.147
   slice methods, 88.147
   two-beam theory, 88.147
Dynamic differential pumping, water vapor, of, 8.75
Dynamic diffraction, 11.89
Dynamic RAM (DRAM)
   bipolar DRAM cells, 86.59–62, 82
   future directions, 86.64–74
   heterostructure DRAM cells, 86.44–58
   integrated circuit technology, 86.2, 4–5
   JFET and MESFET DRAM cells, 86.32–44
   modulation-doped heterostructure DRAMs, 86.54–58
   quantum-well floating-gate DRAMs, 86.49–54
   undoped heterostructure DRAMs, 86.44–49
Dynamic scale-measuring instruments, 8.49

Dynodes:
  schematic of, 13.60, 66
  stack of, in photomultiplier tubes, 13.67–74
  structure of, 13.68–69
Dyson interferometer microscope, 1.68, 80

E

EA-ROMs *See* Electrically altered ROMs
EAXIAL (software), 13.26
  axial fields in, 13.42–44
  and Wien filter, 13.115
E-beam *See* Electron beams
E-beam induced current analysis *See* Electron beam induced current
EBIC *See* Electron beam induced current
EBS tubes, 9.31, 40, 41, 43, 44, 60
EBT *See* Electron beam testing
ECONT (software), 13.26
  computation, high beam energy scan, 13.82
  equipotential plots in, 13.41–42
Eddy current, field, 82.2, 3, 20
  equations, 82.21
    potential formulations, 82.29, 31, 34–36
EDFAs *See* Erbium-doped fiber amplifiers
Edge:
  active, 85.131
  digital image, 93.231
  modeling, 8.321; 88.322
Edge-based analysis, 86.125, 129–130
Edge detection:
  adaptive filters, 88.308
  defined, 88.299
  Gabor functions, 97.7
  group algebra, 94.45
  Hilbert transform pair, 88.308
  linear filtering, 88.310
  model-based, 88.308
  nonlinear filtering, 88.307
  operators, 93.232–239, 259, 320, 323
  optical symbolic substitution, 89.87–89
  orientation analysis, 93.220, 231–239, 300
  performance assessment of, 88.340
  photoelectric microscope, 8.44–46
  presentation of results, 93.239–244
  quadrature filters, 88.308
  region, 88.307
  robust, 88.308
  space-variant image restoration, 99.312–34
  template matching, 88.308
  wavelets, 97.63–64
  zero crossing, 88.332
Edge detection filters:
  Canny, 88.321, 327
  desirable properties of, 88.311
  difference of boxes, 88.311
  Gaussian approximations, 88.326
  infinite impulse response, 88.329
    Deriche, 88.329
    drawbacks of, 88.331
    recursive implementation of, 88.330, 332
    Sarkar and Boyer, 88.329
    Shen and Castan, 88.331
  matched, 88.318
  Petrou and Kittler, 88.322
  quality measure of:
    composite performance, 88.320–321
    false maxima, 88.319
    good locality of, 88.315, 332
    signal-to-noise ratio, 88.312, 332
  scaling of, 88.315, 318
  Sobel, 88.304
  Spacek, 88.321
  spline, 88.339
  zero crossing, 88.332
Edge-enhancement, 10.10, 13
Edge-preserving reconstruction algorithms, 97.91–93, 118–129
  extended GNC algorithm, 97.132–136, 171–175
  generalized expectation–maximization (GEM) algorithm, 97.93, 127–129, 153, 162–166

Edge-preserving reconstruction algorithms, *continued*
  graduated nonconvexity (GNC) algorithm, 97.90, 91, 93, 124–127, 153, 168–175
Edge states, mesoscopic physics, 91.223–224
EDX *See* Energy dispersive X-ray analysis
EELS *See* Electron energy loss spectroscopy
Effective Bohr radius, 82.205, 207, 230
Effective-domain model, 87.134–136, 145–147
Effective Rydberg energy, 82.205
Efficient rational algebra, 90.99–119
EFPLOT (software), 13.42, 43–44
EG&G oscilloscope, 91.234
EGDE *See* Ensemble GDE
Egg white denaturation, 8.113
Eigenfunction, 82.40
Eigenindex, 90.56, 64
Eigennode, 84.112; 90.56
Eigenproblem, 90.60–61
  image algebra, 84.111–114
Eigenspace, 90.68–79
Eigenvalue, 82.4, 40, 43, 45, 49, 62, 63, 86, 87, 363; 85.33; 90.60–63
Eigenvectors:
  equivalent eigenvectors, 90.65–67
  finite eigenvectors, 90.63–67
  fundamental eigenvectors, 90.63–64, 67
  left-hand, 90.252
  right-hand, 90.251–252
8,2 formula, 93.252, 253
8,5 formula, 93.252, 253, 257, 261–262, 264, 266, 269, 277
Eighth-order Hamiltonian function, power-series expansion, 91.9–10, 13
Eikonal aberration theory, 91.1, 2
Eikonal functions, power-series expansion, 91.13–35
Einstein, Albert, 10.270
Einstein–Podolsky–Rosen (EPR) experiment, 94.302–304

Einstein's field equations, information approach, 90.174, 186
Einstein's photoelectric equation, 10.270
Einzel lens, 11.118; 13.134, 138
Elastic collisions, electrons, 89.100–103
Elastic electron scattering:
  crystal disorder, 90.216
  electron diffraction, 90.216–217
  of light, 12.248
  phase problem and, 7.245
Elasticity problems, 82.357
Elastin, negative staining, 6.233
ELEC3D (software), 13.25, 27, 77–78
Electrical bandwidth, defined, 99.76
Electrical current, detection, 87.180–181
Electrically alterable ROMs (EA-ROMs, E-PROMs), 86.71
Electrical supplies, mirror electron microscopy, 4.214
Electric field intensity, 82.4, 10, 26, 36
Electric fields:
  electron-optical properties, 89.401–403
  in electron optical systems, 13.3
  image, effect on, 1.152
  and magnetic fields, combined, 13.107–119
  micro-electric field, 13.76–78
  polarization, chiral media, 92.130–132
  software for, 13.25–44
Electric flux density, 82.4, 19
Electric force microscopy, 87.129–133
  constant-compliance measurement, 87.132
  Derjaguin approximation, 87.130
  electrostatic probe–sample interactions, 87.129–131
  Hamaker constant, 87.132–133
  operational conditions, 87.131–133
  polarizable probe, 87.130–131
  "servo force," 87.129
Electric lens with elliptical defects, tolerance calculations, 13.54–59
Electric vector potential, 82.19, 42
Electric wall, 82.39, 86, 90, 92
Electrochemical transistor, 91.144

Electrochromic display, flat panel technology, 91.251–252
Electrodes, 83.12, 14
  arrangements of, 13.132
  collecting, 83.49, 73
  curved, in FDM methods, 13.8–9
  extraction, 83.43, 47
  gate, 83.30, 43, 47, 49, 78, 80, 85
  interdigital, image, 4.255, 266
  pierce, 83.49, 73
  representation of, 13.32–33
Electrodynamics:
  information approach, 90.149–153
  Maxwell's equations, 90.186–187
  from vector equation, 90.196–198
Electroluminescence, pulsed dc, 91.245
Electrolysis, 83.27, 28
Electrolyte solution, effect on ionic forces, 87.106–112
Electromagnetic circulators *See* Circulators
Electromagnetic coupling:
  coupling coefficient, 89.213–214
  optical interconnects, 89.210–213
  quantum devices, 89.208–243
Electromagnetic crosstalk, quantum devices, 89.209
Electromagnetic field:
  distributions, computing methods in, 13.4–6
  highly anisotropic media, 92.80–200
  misalignment caused by, 83.223–224, 230
  parasitic aberrations, 86.226, 234, 235, 245, 261
  Schrödinger equation and solution in, 13.248–251
  stray, 83.221
  transverse, SE collection, 83.223, 230
Electromagnetic lenses, 2.5; 10.234–236; 91.261–262, 264–265; 93.187–188
  aberration reduction, 8.246–246, 256
  assessment, 5.233, 234
  axial field computation in, 13.51–54
  axially symmetric lenses, 97.282–316, 333–335

canonical aberrations, 97.381–388
charged-particle wave optics:
  axially symmetric lenses, 97.282–316, 333–335
  quadrupole lenses, 97.317–320, 336
  circular loop in, 13.140–141
  design, 2.225; 5.209, 210
  with elliptical defects, 13.57–59
  tolerance calculations, 13.54–59
  FEM and FDM on, 13.129, 153
  field distribution, 1.125, 143
  field measurement, 2.228
  flux density, 5.212–214
  focusing:
    first-order focusing, 13.182–185
    in scanning electron microscope, 13.113–114
  geometry, 13.52
  Glaser function, 5.213
  Glaser model, 5.212–214
  IEM on, 13.153–154
  integration transformation, 97.369–381, 389–392
  lens field distribution, 1.125, 143
  lens fields, 13.140–145
  limitations, 5.209
  multislice approach, 93.179, 181–182, 187
    cylindrical, 93.175–176
    Glaser–Schiske diffraction integral, 93.195–202
    improved phase-object approximation, 93.190–192
    paraxial properties, 93.194–207
    quadrupole, 93.181–185
    spherical aberration, 93.207–215
    spherical wave propagation, 93.194–195
    thick lens theory, 93.192–194
  objective lens design, 5.211–214
  optical properties, 83.137
  optimization, 13.207–216, 209–210
  power-series expansions:
    eikonal, 97.366–369
    Hamiltonian function, 97.361–366

Electromagnetic lenses, *continued*
  round symmetric, 93.186–202
  spherical aberration, 1.216
  superconducting, 5.210
Electromagnetic properties, highly anisotropic media, 92.80–200
Electromagnetic scanners, 8.36
Electromagnetic wave, 82.3, 38
Electromechanical switching system, 91.192, 201–202
Electromigration, 83.16, 63; 87.208, 221, 241
Electron *See* Electrons
Electron beam *See* Electron beams; Electron beam testing
Electron beam holography *See* Holography
Electron beam induced current (EBIC), 10.61–65; 12.150; 13.211
Electron beam inspection system (EBIS), 13.124–125, 179–180
Electron beam lithography, 13.179, 188
Electron beam microanalysis, 9.299, 316, 317
Electron beams, 83.3, 113
  brightness, 8.139, 171, 173, 177–179, 181–185, 248–251
  energy loss in the specimen, 1.124
  exposure, linewidth control, 83.186–192
  highly directional, 8.137
  lithography, 83.36
  metallization, 86.30
  monochromatic beam and nanotip, 95.115, 118
  nanometer, 83.116
  vector scan, 83.113
Electron beam testing (EBT), 12.109, 110, 141; 13.124–125
  deflection elements in, 13.179–180
  development of, 13.126–127
  and electrostatic charging, 13.230
  and magnetic lenses, 13.145
  planar multipole representation in, 13.130–135
  probe fields in, 13.153

Electron biprism, 12.5, 11, 16, 17; 91.282; 99.173
  image wavefunction and, 99.179–184
  wave-optical analysis, 99.176–179
Electron diffraction, 2.197, 203, 212, 241; 4.113–114; 90.206–350
  of ash minerals, 3.135–138, 145
  selected area, 3.138, 161, 173
  small-angle techniques, 3.155–218
  background, 90.206–207
  Bloch wave channeling, 90.293–334
  camera length, 88.123
  convergent-beam *See* Convergent-beam electron diffraction
  crystal structure factors:
    direct inversion, 90.257–265
    potential and, 90.338–341
  diffracted beam amplitude, 90.209
  diffraction geometries, 88.123
  diffraction patterns, 90.209–214
    recording, 88.122
  dynamical elastic diffraction, 90.221–250
  dynamical theory, 93.58, 60
  goniometry, 88.124
  Green's functions, 90.334–338
  illumination of sample, 88.122
  imaging plate, 99.270–274
  intensity data, 88.125
  perturbation methods:
    nonperiodic structures, 90.272–293
    periodic structures, 90.250–272
  phase grating theory, 5.249
  phase problem solution and, 7.220–225
  potential, 90.251
    crystal structure factors, 90.338–341
    full potential mode, 90.245, 248
    optical potential, 90.216–217, 341–349
    truncated potential mode, 90.244–245
  reflection high-energy electron diffraction (RHEED) *See* Reflection high-energy electron diffraction
  scattering:
    axial resonance scattering, 90.302–310

by average potential, 90.214–216
diffuse scattering, 90.217–218
elastic scattering, 90.216–217
planar resonance scattering,
  90.313–317
quasi-elastic scattering, 90.217–218
real diffuse scattering, 90.348–349
resonance scattering, 90.293–334
selvage scattering, 90.241, 245–248
substrate scattering, 90.241, 243–245
surface resonance scattering,
  90.323–334
TDS scattering, 90.286–287
virtual diffuse scattering, 90.348–349
scattering amplitude, 90.207–209
scattering cross-section, 90.207–209
small-angle diffraction, 7.163;
  98.358–360
theory, 90.207–221
weak phase approximation and,
  7.199–211
Electron diffraction camera:
commercial development in Europe,
  96.427, 441, 488–489, 500–501,
  519–520
Japanese contributions, 96.599–600
Electron diffraction spectroscopy, 11.94
Electron distortion field, 94.109–114
Electron emission *See* Field emission
Electron energy loss spectroscopy (EELS),
  9.65–171; 11.44, 77, 89
analyzing mode, 9.69, 130
background contribution, 9.76, 124–127
band structure effects:
  ELNES and, 9.119–120
  solid-state environment and,
    9.113–114
beam-induced damage, minimum
  detectable mass and, 9.159–160
Bethe theory and, 9.95–111
bidimensional pictures, 9.69
biological applications, 9.150
carbon:
  K-edge extended fine structures,
    9.111–113, 123–124

K-edge in nucleic acid bases, 9.113,
  116, 163
chemical mapping, energy filtered
  images, 9.163–166
chemical shift:
  edge shape and, 9.114–119
  oxidation state and, 9.115, 116,
    118
  solid-state threshold and, 9.111, 113,
    162
classification of systems, 9.131–132
core-loss excitation, 9.67, 95, 127–130
coupling mode, 9.68, 69
cross-section, quantitative microanalysis
  and, 9.154–155
CTEM, energy filtered images, 9.130,
  131–132, 165
data processing, 9.145–146
detection limits, 9.158–162
detection unit, 9.141–145
diffraction conditions, quantitative
  microanalysis and, 9.157–158
earth sciences applications, 9.150
edges:
  library of, 9.147–150
  shape classification, 9.102, 106–107
  $2p$ edges in transition metals, 9.116
  valence state of excited atoms, 9.95
elastic scattering and, 9.73–74
elemental analysis and, 9.146
elemental concentrations, 9.146
elementary excitations and, 9.75–82
energy filtered images and, 9.169–170
energy losses on surfaces at glancing
  incidence, 9.170
environmental information, 9.162–163
experimental parameters, 9.69–70
extended fine structures (EXELFS),
  9.111–113, 114, 120–124, 162–163
filtering mode, 9.69, 130
hematin, 9.75, 76
high-energy loss region, 9.75, 94–130
high-voltage microscopy and, 9.166,
  168–169
historical development, 9.66–67

Electron energy loss spectroscopy
(EELS), *continued*
 inelastically scattered electrons, angular
  dependence of, 9.71–72, 91–94
 inelastic scattering, 9.68–82
  by valence electrons, total
   cross-section for, 9.93–94
 instrument, 9.130–146; 11.44
 intermediate energy loss domain,
  9.124–130
 ionization edge, specimen composition
  and, 9.95
 low-energy loss region, 9.66, 75, 82–94
 macrophage vacuole, iron distribution
  in, 9.166
 material sciences applications, 9.150
 microanalytical application, 9.69,
  146–166
 minimum detectable mass (MDM),
  9.158–160
 minimum detectable mass fraction
  (MDMF), 9.160–161
 multiple inelastic scattering, specimen
  thickness and, 9.76–77
 near-edge fine structures (ELNES),
  9.113–114, 120, 162–163
 oxygen, K near-edge fine structure,
  9.118
 parallel detection systems, 9.143–145
 partially integrated cross-sections,
  9.108–111
 photodiode arrays, 9.143, 144
 plasmon peak position, q dependence,
  9.93–94
 post-specimen lenses (PSL), 9.140
 qualitative microanalysis, 9.147–150
 quantitative microanalysis, 9.67,
  150–158
 rare-earth oxide thin film, 9.75, 76
 scanning transmission electron
  microscopy, 9.130, 132–133, 140
  energy filtered images, 9.165
 semiconductor detectors, 9.143–145
 single-electron excitations, 9.92–93
 single loss spectra, 9.143
 recovery, 9.76–82
 solid-state environment and, 9.111–114,
  162
 solid thin films, 9.146–147
 specimen thickness, quantitative
  microanalysis and, 9.156
 spectrometer:
  coupling to microscope column,
   9.137–141
  design parameters, 9.134–137
  working mode, 9.68–69
 surface barrier solid-state detector, 9.143
 threshold shape, 9.162
 yttrium, 9.87–88, 127
 yttrium sesquioxide, 9.88–91, 127
Electron gas energy distribution, 8.209
Electron gun beams:
 aberrations, 8.144–147
 accelerating field, 8.139
  energy distribution, 8.138–139,
   186–204
 angle-dependent current density, 8.171
 Boersch effect, 8.178, 189, 220, 256
 brightness, 8.139, 166–186
 cathode surface poisoning, 8.179
 cathode temperature measurement,
  8.178
 cathode tips, 8.178, 183–186
 crossover position and source,
  8.140–144
 current density distribution in crossover
  plane, 8.173
 emission current density, 8.167, 178, 180
 emittance diagrams, 8.152–153,
  157–162
 energy characteristics, 8.166
 energy distribution, 200–203; 8.138–139
  aperture, effect of, 8.186–188
  asymmetry, 8.195
  axial beamlets, 8.193–197
  Boersch effect, 8.178, 189
  current density, 8.194–195, 197
  energy half-width, 8.194
  experimental determination,
   8.190–193

off-axis beamlets, 8.192–193,
  197–199, 199–200
total beam, 8.199, 200–203
geometry, 8.138, 147–166, 153–157
imaging field, 8.139–140
model, 8.139
paraxial model, 8.140–141
pencil of rays, 8.139–140
production, 8.137
ray characteristics, 8.155–157
space charge effects, 8.199
thermionic cathodes with, 8.166–168
total emission current density, 8.167
Electron guns, 10.244–250; 83.6, 35,
  75–77, 125; *See also* Electron gun
  beams
alignment, 83.127
brightness, 1.135; 11.3–6, 104
field emission, 10.248
lanthanum boride, 10.248, 250
paraxial model, 8.140–142
telefocus, 10.247
thermionic cathodes, 10.244–248
triode, 10.248
tungsten hairpin, 10.248–249
"Wehnelt" electrode, 10.245–247
Electron hole pairs, 83.21, 28, 29
Electron holography, 7.164–169; 12.34;
  98.331, 362–363, 422–423;
  99.171–173, 235–236
  Aharonov–Bohm effect, 99.172,
    187–192
  charged dielectric spheres, 99.174,
    207–216
  charged microtips, 99.229–235
  double-exposure electron holography,
    99.205–207, 207–210
  electron biprism, 99.173
    image wavefuction and,
      99.179–184
    wave-optical analysis, 99.176–179
  electron–specimen interaction,
    99.174–176
  electrostatic Aharonov–Bohm effect,
    99.172, 187–192

off-axis electron holography,
  98.323–324
phase-object approximation, 99.174,
  175–176, 184–186
principles, 98.363–373
reverse-biased p-n junctions,
  99.172–173, 174, 185, 216–229
scanning transmission electron
  microscope holography,
  98.373–387, 422–423
Electronic displays, 91.231–256
  cathode ray tube, 91.231–233, 235–238,
    240, 253
  flat panel technology, 91.244–252
    light controller displays, 91.246–247
    light emitters, 91.245–246
    liquid crystals with memory,
      91.247–252
  frame store, 91.242, 244
  graphic display, 91.234–236, 237, 244
  oscilloscope, 91.233–234
  projection display, 91.252–253
  storage technology, 91.237–238
  three-dimensional display, 91.252–253
  vacuum fluorescent display, 91.236
  virtual reality, 91.208–209, 244,
    255–256
  visual display unit, 91.238–242, 243
Electronic newspaper, 91.209
Electronic permittivity, electrostatic limits,
  87.80
Electronics industry, mirror electron
  microscopy, 94.96–98
Electronic switching system, 91.202
Electronic viewing screens, 10.251–252
Electronic viewing systems, 11.9–11
Electron image analysis, 10.9
Electron image data, phase problem
  solution and, 7.220–225
Electron impact ionizer, 83.3
Electron interference, 7.164–169
Electron interferometer, 12.7, 12, 20, 21
Electron lenses, 11.6–8; 83.207, 256
  aberrations, 83.207, 221, 222, 228, 229,
    230

Electron lenses, *continued*
  chromatic aberration, 2.189, 193
  diffraction aberration, 2.172
  spherical aberration, 2.171
  characteristics, 2.229
  condenser-objective, 2.174
  cylindrical, 93.175–176
  deceleration, 83.227
  design, 2.225
  developments, 83.222
  electrostatic, 10.241–244
    immersion, 10.244
    objective, 10.242–244
  instability, 1.153
  for low voltage SEM, 83.222
  magnetic, 2.5; 10.234–236, 234–241; 93.187–188
    aberration reduction, 8.246–246, 256
    assessment, 5.233, 234
    axial field computation, 13.51–54
    canonical aberrations, 97.381–388
    charged-particle wave optics, 97.282–320, 333–336
    condenser-objective, 10.234–246
    design, 2.225
    with elliptical defects, 13.54–59
    FEM and FDM on, 13.129, 153
    field measurement, 2.228
    first-order focusing, 13.182–185
    flux density, 5.212–214
    focusing, 13.113–114, 182–185
    geometry, 13.52
    Glaser function, 5.213
    Glaser model, 5.212–214
    IEM on, 13.153–154
    integration transformation, 97.369–381, 389–392
    lens fields, 13.140–145
    limitations, 5.209
    objective lens design, 5.211–214
    optical properties, 83.137
    optimization, 13.207–216
    power-series expansions, 97.361–369
    principles of design, 5.209, 210
    second-zone, 10.236–237
    single-pole, 10.237–238
    superconducting, 5.210; 10.238–241
  objective, 2.224
  paraxial properties, 93.194–207
  quadrupole, 93.181–185
  resolution parameter, 2.173
  round symmetric, 93.186–202
  stability, 2.190
  trapped flux electron lenses, 5.220–224; 14.19–20
Electron metrology, 12.154
Electron microdiffraction, 2.197
Electron micrograph:
  calibration, 6.255
  image reconstruction, 6.252, 253, 255–258
  image superimposition, 6.243
  integration, 6.243, 245, 246, 247, 249, 250, 253
  linear periodicity integration, 6.246, 249, 250
  with nonrepeating features, 6.269
  optical diffraction analysis, 6.249, 250, 251, 252, 253, 254
  optical reconstruction, 7.253–256
  repeating feature reinforcement, 6.243, 245, 249
  with repeating features, 6.243, 253, 254, 257, 269
  two-dimensional lattice integration, 6.243, 244
Electron micrograph diffraction pattern:
  diffraction spot filtering, 6.257
  formation, 6.249, 250, 251, 252, 253, 254
  information, 6.254
  masking, 6.255, 257, 258
  micrograph quality and, 6.255
  noise separation, 6.257
  noise spectrum, 6.254, 257
  repeating feature representation, 6.257
  resolution, 6.254
  underfocusing, 6.254
  of viruses, 6.253, 254, 256, 265, 266

Electron microprobe:
commercial development in Europe, 96.460–461, 464–466
Southern Africa research, 96.341–342
use, 5.41
Electron Microscope Society of America (EMSA)
formation, 96.4, 357
growth, 96.358–360
international relationships, 96.358–359, 364, 368
meetings, 96.357–358, 374–375
presidents, 96.357, 374–375
publications, 96.359–361, 851–875
Electron microscopy, 2.167; 5.163, 165, 167, 187, 211, 214, 217, 244, 245, 287, 288, 298; 89.2, 4, 6; 91.259–289; 94.197
aberrations, 10.230–232
chromatic aberrations, 5.181, 298
spherical aberrations, 5.196, 197, 298
of acid phosphatase, 6.180, 181, 182
of adenyl cyclase, 6.173, 174, 184–188
of alkaline phosphatase, 6.173, 174, 184, 185, 186, 187, 188
amplitude contrast, 5.176–178, 187–193
analytical, 10.257–259
aperture, 5.165
approximations:
first-order, 5.169–183
for interpretation, 5.167, 168
second-order, 5.183–195
of ATPase, 6.191, 192, 193, 194
in azo dye technique, 6.186, 187, 188
of biological materials, 2.242
at low temperatures, 5.310
Brownian notion effect, 5.339, 340
by strioscopy, 5.300, 306, 320–328, 334, 341, 342
dark-field technique, 5.299, 300, 302, 303
diffraction, 5.299
image enhancement, 5.309, 310
intrinsic contrast, 5.299
of lipase, 6.198
of lysosome, 6.183
phase contrast technique, 5.299
resolution, 5.196, 298–303
of viruses, 6.227, 228, 229, 230, 233, 239, 240, 241, 242, 243, 244, 245, 246, 261
Castaing and Henry analyzer, 10.320
cathode ray oscilloscopes, 10.218–221
of cholinesterase, 6.200, 202, 203
contrast in, 1.129
crystal-aperture scanning electron microscopy, 93.57–107
experimental, 93.66–87, 90–91
imaging, 93.58–59, 63–66, 87–90, 94–106
resolution, 93.91–94
theory, 93.59–66
crystal disorder, 7.288
in DAB technique, 6.210
defocusing, 5.163, 165, 174, 179, 196–198, 240, 241, 245, 249, 294
of dehydrogenase, 6.209, 214
diagram, 10.217
diffraction electron microscope, 91.278–279
distribution artifacts, 8.114–115
electron guns, 10.244–250
electron mirror, 10.229
electrostatic electron microscope, 91.274, 278
electrostatic lenses, 10.224–226, 241–244, 325–326
elementary image, 1.128, 133
element discrimination, 5.180
emission, 10.227–228, 246
energy analyzing microscope, 4.318–340
energy analyzing and selecting devices, 4.263–318
energy selecting microscopy, 4.340–358
energy spread, 7.135
of enzymes, 6.172, 174, 175, 176, 177, 178, 179, 204, 219, 220, 221, 222
errors of, 1.120

Electron microscopy, *continued*
  of esterase, 6.187, 188, 195, 196, 197, 198, 199, 206
  field emission guns, 10.248–250
  fluorescence microspectroscopy and, 14.172–174
  focusing error, 1.125, 132
  high-resolution electron microscopy *See* High-resolution electron microscopy
  high tension, 1.146
  high-voltage electron microscopy *See* High-voltage electron microscopy
  history, 10.215–218
    Belgium, 10.221–222
    Britain, 10.221–224, 254–256
    Germany, 10.218–221
    Holland, 10.222–223
  image formation, 5.164–167; 10.229–233
  image intensification, 5.309, 340, 343
  image reconstruction, 5.197, 198
  in-focus image, 5.180
  lattice images, 7.163
  lenses; *See also* Electron lenses
    condenser-objective lens, 10.234–236
    cryo-objective lens, 10.239–241
    immersion lens, 10.244
    magnetic, 10.234–236
    second-zone lenses, 10.236–237
    single pole lenses, 10.237–238
    superconducting, 5.202, 215–235; 10.238–241; 14.24–25
    types, 10.234–244
  linear energy transfer value, 5.311, 312
  low-energy electron microscope, 94.86
  low-magnification, diffraction-contrast, 88.126
  low-temperature, 5.310
  macromolecular three-dimensional structures from, 7.281–377
  magnetic *See* Magnetic electron microscopy
  magnetic microstructure, 98.330–333
  magnification, 6.253
  membranes, 7.259, 288
  of metals, 2.231
  metioscope, 10.228
  mirror electron microscopy *See* Mirror electron microscopy
  Mulvey's projector lens, 10.239
  observation and recording, 10.251–253
    with optical diffractometer, 6.250–255
  optics, 93.174–176, 215, 216
  origins, 10.218–229
  of oxidation/reduction enzymes, 6.209, 210, 214, 215, 216
  partial coherence, 5.182, 183
  patent, 96.134
  performance, 6.251
  permissible disturbances in, 1.124
  phase contrast, 1.172, 178; 5.169–176, 193–195
  phase contrast transfer function, 88.129
  phase problem in, 7.186–279
  Philips Analytical, 10.258–259
  photoelectron microscopy *See* Photoelectron microscopy
  photographic averaging and, 6.243
  photography, 10.252–253
  radiation damage, effect of, 88.156
  radiobiological data, comparison with, 5.311, 312, 313
  reflection type, 4.163; 10.228–229
  replicas, 5.40
  residual gas in, 1.157
  resolution, 5.173, 175, 176, 177, 188, 190, 191, 192, 211, 215, 233, 234, 240, 241–243, 298–303, 343; 6.251
  resolution limit, 1.115
  sample preparation, 8.119–127
  scanning electron microscopy *See* Scanning electron microscopy
  scanning transmission electron microscopy *See* Scanning transmission electron microscopy:
  shadow-electron type, 10.229
  shrouded coils, 14.21–23
  signal-to-noise ratio, 5.309, 310

specimen:
  cooling, 1.164–171
  ionization, 1.157
  thickness effect, 5.180, 181
specimen grids, 3.104
stereo examination, 5.339, 340
of sulfatase, 6.181, 182, 183
superconductors in, 14.16–25
  diamagnetic shielding lenses, 14.20–21
  early work, 14.16–17
  shrouded coils and pole piece lenses, 14.21–23
  solenoid lenses, 14.18–19
  trapped-flux lenses, 14.19–20
in tetrazolium salt technique, 6.214
thermionic cathodes, 10.244–248
thin films, 5.40, 41
total image, 5.178
of tracer enzymes, 6.216, 217, 218
transmission electron microscopy *See* Transmission electron microscopy
two-stage imaging, 10.224
ultra-high voltage, 11.59, 75, 76
ultra-short exposure, 5.340
underfocused image, 5.180, 196
viewing screens, 10.251–252
"Wehnelt" electrode, 10.245–247
Electron mirror, 12.45
Electron mirror microscopy, 10.229
Electron mirror system (EMS), 1.229, 230
  contrast mirror electron microscopy, 94.108–109
  mirror electron microscopy, 94.87–88, 99–108
Electron multiplier, 2.9, 10, 23, 24
Electron off-axis holography, 7.250; 12.33; 89.1–48, 4, 5; 94.199; 98.323–324
  applications, 89.36–47
  crystal defects, 89.44–47
  dynamical phase effects, 89.38–43
  thickness measurement, 89.36–38
  digital reconstruction, 89.5–6, 12–18, 21–24, 33–34, 36, 47

phase distribution, 89.18–19, 41
problems, 89.25–35
  hologram recording, 89.32–35
  limited coherence, 89.25–31
  noise problems, 89.31–32
reconstruction:
  digital, 89.5–6, 12–18, 21–24, 33–34, 36, 47
  light optical, 89.5–6, 10–12, 21–24, 33–34
scanning transmission electron microscopy, 94.232–244, 252–253
transmission electron microscopy, 94.253–256
Fresnel diffraction, 94.253–256
Electron optical systems; *See also* Electron optics
  aberrations in, 13.189–199
  design aims, 13.3–4
  electromagnetic field distributions in, 13.4–6
  Fourier–Bessel series in, 13.135–136
  inspection, low-voltage, 13.127–128
  machining tolerances, calculation, 86.225–279
  magnetic microstructure, 98.331–332
  modeling, 13.1–3
    relaxation methods, 13.6
  numerical modeling *See* Numerical modeling
  parasitic aberrations, calculation, 86.225–279
  planar multipole representation, 13.130–135
  probe fields in, 13.153
Electron optical transfer function:
  and axial astigmatism, 13.278–279
  basic concepts, 13.272–274
  and rotationally symmetric third-order aberrations, 13.280–284
  and spherical aberration and defocus, 13.274–277
Electron optics, 12.149; 83.222; 93.64, 174–176, 187–190; *See also* Electron optical systems

Electron optics, *continued*
  Aharonov–Bohm effect, 93.174, 176, 178–181
  canonical aberration theory, 97.360–406
    to ultrahigh-order approximations, 91.1–35
  coherence in, 7.101–184
  Glaser–Schiske diffraction integral, 93.195–202
  history at international congresses, 96.405–412
  improved phase-object approximation, 93.190–192
  lenses:
    axially symmetric electrostatic lenses, 97.320–321
    axially symmetric magnetic lenses, 97.282–316, 333–335
    magnetic quadrupole lenses, 97.317–320, 336
  multislice approach, spherical aberration, 93.207–215
  paraxial properties, 93.194–207
  spherical wave propagation, 93.194–195
  thick lens theory, 93.192–194
Electron pairing, in high-temperature superconductivity, 14.38–39
Electron probe, 83.205
Electron probe microanalysis, 3.146, 152; 13.126
  absorption effect, 6.287, 288
  accuracy, 6.276, 279
  area scanning with ratemeter, 6.294–296
  atomic number effect, 6.287, 288
  background, 6.286, 287
  beam excursion, 6.279, 281
  biological material, 6.282
  complete area scan, 6.277
  deadtime, 6.285, 286
  display system, 6.278
  electron beam stability, 6.284
  electron diffusion, 6.282
  in element analysis, 6.275
  energy-dispersive solid state detector, 6.280, 281
  field aberrations, 6.282
  fluorescence effect, 6.287, 288
  image recording, 6.278
  linear scanning, 6.296, 297
  magnification, 6.278, 279
  matrix scanning, 6.298, 299
  mechanical scan, 6.282
  nondestructive analysis, 6.276
  probe size and aberrations, 13.195–199, 292–295
  qualitative, 6.275, 277, 281
  quantitative, 6.275, 277, 280, 281, 287
  quantitative area scanning, 6.298, 299
    instability, 6.284, 285
  ratemeter signal manipulation, 6.297
  scanning procedure, 6.277, 278, 297
  secondary X-ray production, 6.282
  semi-focusing system, 6.280, 281
  sharpness of image, 6.279
  signal intensity, 6.282
  signal intensity loss, 6.279
  sources of instability, 6.284, 285
  spatial resolution, 6.278, 281, 282
  spatial selectivity, 6.276, 277
  specimens:
    displacement, 6.281
    movement, 6.282
  spectrometer defocusing, 6.280, 281, 282, 297
  standard X-ray area scan, 6.288–292
  width of scan, 6.278, 279, 280
  X-ray excitation, 6.275, 276, 277
  X-ray intensity measurement, 6.283, 284
  X-ray line choice, 6.282
  X-ray line interference, 6.281
  X-ray scan color composites, 6.292–294
Electron ray model, 93.60–61
Electron ray simulation, predictions, zone axis pattern, 93.66–73
Electrons; *See also* Electron optical systems; Electron optics; Electron wave devices; Electron wave optics:
  affinity, 83.10, 20, 21, 24, 28
  backscattered, 83.205–206
  bombardment, 83.34, 65

INDEX

charge:
  conservation of, 13.254–255
  effect of, 83.208
cloud, 83.22
collection efficiency, 13.82–85, 89–91
  in energy scans, 13.82–85, 89–91
  in low-voltage inspection, 13.128
  in SEM techniques, 13.75–76
collisions, 83.19, 21, 63
  elastic and inelastic, 83.208
density exposure curves, 1.188
diffraction, dynamical theory, 93.58, 60
disadvantages, 83.206
drift diffusion formalism, 89.110–113
elastic-inelastic collisions, 89.100–103
electron–electron interaction, 82.199
electron–impurity interaction, 82.199; 92.223
energy, 83.7
high energy, 2.169, 176
inelastic collisions, 89.100–103
as microscopic probes, 83.206–220
noise, 1.195
particle behavior, 89.110–111, 118
particle–wave duality, 98.333–335
penetrating power, 2.176
  in crystals, 2.182
primary *See* Primary electrons
quantum-coupled spin-polarized
    single-electron logic devices, 89.223–235
range, 83.218
range in emulsions, 1.192
scanning, 83.205
scattering *See* Electron scattering
secondary *See* Secondary electrons
sources of *See* Electron sources
specimen, effect on, 1.154
spin *See* Electron spin
trajectories *See* Electron trajectories
transmission through crystal lattice, 93.59–62, 66–73
transport *See* Electron transport
wave behavior, 89.103–110, 118
wavelength, 2.170; 83.205, 207, 221

Electron scattering, 2.177 et seq., 11.19; 12.26; 83.120, 208, 217–219, 235
  anisotropic, 84.333
  axial resonance scattering, 90.302–310
  Bloch waves:
    axial resonance scattering, 90.302–310
    planar resonance scattering, 90.313–317
  Bragg diffraction, 84.322
  carrier-impurity, 82.200
  combined, model, 84.332
  defects diffuse scattering, 90.282–285
  diffuse, 90.217–218, 279–285
    model, 84.329, 332, 334
  elastic *See* Elastic electron scattering
  electron diffraction, 90.214–218
  impurity, 82.208, 211, 216, 227
  inelastic *See* Inelastic electron scattering
  intervalley, 82.199, 200, 216, 217, 269
  isotropic, 84.332
  Monte Carlo simulations, 83.217–218, 241–242
  multiple, 82.200, 242
  planar resonance scattering, 90.313–317
  plural, 2.179, 210
  Rayleigh, 84.330, 332
  real diffuse scattering, 90.348–349
  resolved structure, 84.333, 334, 336
  selvage scattering, 90.241, 245–248
  small angle theory, 3.206–214
  structural, model, 84.332
  subresolution structure, 84.333
  substrate scattering, 90.241, 243–245
  surface resonance scattering, 90.323–334
  TDS scattering, 90.286–287
  unresolved structure, 84.336
  virtual diffuse scattering, 90.348–349
Electron scattering factor, 86.189, 191, 221
Electron sources, 83.6, 31, 32, 207
  addressable, 83.35
  brightness, Langmuir, 83.221–222
  coherence, 7.106

Electron sources, *continued*
  energy spread, 83.7, 43, 55, 57, 75, 76, 206, 208, 225, 229
  field emission, 83.222, 228–229, 256–257
  ion, 83.82
  lanthanum hexaboride, 83.222
  liquid metal, 83.32
  low voltage, 83.221, 257
  Schottky, 83.229
  silicon, 83.33
Electron spectroscopy (XPS), 10.319–321
Electron spin, electron optical coherence and, 7.172
Electron trajectories, 13.173–216; 83.61
  aberrations in, 13.189–199
  deflection, first-order, 13.186–188
  displacement effect in coulomb interaction, 13.220–222
  equation, 13.174–176; 91.4, 13
  first-order or paraxial optics, 13.179–188
  focusing, first-order, 13.181–186
  fundamental laws in, 13.176–178
  optimization, 13.207–216
  ray tracing and, 13.199–207
  reducing displacement effect, 13.222–224
  Schrödinger equation and, 13.249
Electron transport, 89.103–119
  ballistic, 89.107
  diffusive, 89.107
  drift diffusion formalism, 89.110–113
  electron wave devices, 89.103–110
  quasi-dissipative, 89.118–119
Electron tunneling, 87.191
Electron wave devices, 89.99–120
  Aharonov–Bohm devices, 89.142–178
  current and conductance formulas, 89.113–118
  directional couplers, 89.193–199
  electron transport, 89.99–113
  energy dissipation, 89.119–120
  potential drop, 89.119–120
  quasi-dissipative transport, 89.118–119
  resonant tunneling devices, 89.123–124

T-structure transistors, 89.178–193
Electron wave directional couplers, 89.193–199
Electron wave function, 90.216
Electron wave guides, 89.193
Electron wave optics, 13.243–301; 93.60–61; *See also* Charged-particle wave optics
  angular divergence and, 13.288–292
  applications, 13.292–298
  chromatic aberration and, 13.285–288
  computer image processing, developments in, 13.245–246
  electron optical transfer function, 13.272–274
  Fraunhofer diffraction, 13.261–268
  high-resolution microscopy, developments in, 13.244–245
  ideal imaging, proof of, 13.256–257
  image formation, theory of, 13.268–285
  wave functions for rotationally symmetric fields, 13.247–255
  in paraxial condition, 13.255–261
Electro-optic light modulator, 1.89, 90, 103; 6.136; 89.199–200
Electro-optic switch, T-structure transistor, 89.191–193
Electrophoretic display, flat panel technology, 91.252
Electrostatic Aharonov–Bohm effect, 89.145–146; 99.187–192
  disordered structures, 89.165–178
  double quantum wells, 89.157–165
  interferometer, 89.172–178, 180, 186, 199
Electrostatic anode lens, field electron emission, 8.242–243
Electrostatic biprism, 99.188–190, 193
Electrostatic charging, 3.193–196
  and Coulomb interaction, 13.224–231
  Boersch effect, 13.216–220
Electrostatic deflectors, 13.146–149
Electrostatic electron microscope (EES), 10.224–226, 241–244; 91.274, 278
  commercial production, 96.141

development, 96.137–141
  at Japanese universities, 96.247–248,
    251–256, 263–267
  origins, 10.219, 222
Electrostatic energy analyzers, mirror
    bank, 89.391–478
  charged particle focusing, 89.399–410
  charged particle trajectory equations,
    89.393–399
Electrostatic field, 82.4, 6, 10
  analytical calculation, 13.128–153
  numerical calculation, 13.153–173
Electrostatic lenses, 10.241–244; 83.62
  aberrations, 8.246
  by synthesis, 13.209–210
  charged-particle wave optics,
    97.320–321
  Einzel lens, 96.139
  with elliptical defects, tolerance
    calculations, 13.55–57
  FEM and FDM on, 13.129, 153
  first-order focusing, 13.181, 182–185
  history and development, 10.242–244
  IEM on, 13.153–154
  integration transformation, 97.396–403
  Japanese contributions, 96.685–688,
    705–707
  lens fields of, 13.137–140
  multislice approach, 93.177–178,
    182–184, 187–188
  optimization, 13.207–216
  patent, 96.137
  spherical aberration, 1.219
Electrostatic multipole, 85.241–245
  M function, 85.241, 243–245
  symmetry transformations, 85.245
  transformation, 85.241–243
Electrostatic point projection microscope,
    10.325–326
Electrostatic potential:
  canonical aberration theory, to
    tenth-order approximation, 91.5
  surface, 87.106–107
Elementary excitations, 9.75–82
  general classification, 9.75–76
  recovery of single loss spectra,
    9.76–82
Elementary path, 90.39, 40
Elementary state functions, 94.329–333
Ellipses, transforms, 7.26
Elliptical defects in lenses:
  in bore, 13.57–59
  tolerance calculations, 13.54–59
Ellipticity, 86.239, 245, 246, 250, 260
Elmsikop, cooling chamber for, 1.161
Embedding, 2.252, 317 et seq.
  with epoxy resins, 2.322
  with methacrylates, 2.318, 324
  with polyesters, 2.320
  with water-soluble resins, 2.323
Embedding media, 3.107, 143, 150;
    96.98–99
Emission electron microscope
  commercial development in Europe,
    96.486, 488
  described, 4.162–163
  diagram, 10.246
  history, 10.227–228
Emission lens, 11.122
Emission spectra, in fluorescence
    microspectroscopy, 14.127–132
Emittance diagrams:
  calculated and measured, comparison of,
    8.157–162
  construction from shadow curves,
    8.152–153
  emission energy, dependence on,
    8.160–161
  evaluation, 8.162, 163–165
Emitters, 83.3
  cones, 83.36–38, 45, 56
  etched wire, 83.47, 53
    results, 83.55
  lifetimes, 83.13
  silicon, 83.3, 24, 46, 47
  wedge, rim and edge, 83.38–42, 43, 46,
    75, 76, 78
  whiskers, 83.54, 55
EMOPAP *See* Extrapolated method of
    parallel approximate projections

EMOPNO *See* Extrapolated method of
  parallel nonexpansive operators
EMOPP *See* Extrapolated method of
  parallel projections
EMOPSP *See* Extrapolated method of
  parallel subgradient projections
EMS *See* Electron mirror system
EMSA *See* Electron Microscope Society of
  America
Emulsification reaction kinetics, wet
  replication study of, 8.99
Encoder, digital coding, 97.192
Endoplasmic reticulum, 6.183, 220
  fluorescent probe of, 14.147–150
Energy absorption spectrum, 87.72
Energy analyzers, 4.263–360
  mirror bank *See* Mirror-bank energy
    analyzers
  two-plate electrodes, 89.392, 410–432,
    441–476
Energy analyzing microscopes, 4.318–340
Energy bands, anisotropic, 92.85–88
Energy dispersive X-ray analysis (EDX),
  11.48; 13.211
Energy eigenfunction, 90.216
Energy filtering, 11.119
Energy loss, 2.192, 193
Energy loss spectrum, 4.264–266,
  329–340, 353–358
Energy–mass relation, quantum
  mechanics, 90.159–160, 187
Energy selecting microscopes, 4.340–358
Energy spread, 11.5, 106
  electron sources, 83.7, 43, 55, 57, 75,
    76, 206, 208, 225, 229
  estimation in electron microscope, 7.135
  of incident beam, 7.160
Energy–time relation, uncertainty
  principle, 90.167–168, 187
England *See* United Kingdom
Ensemble GDE (EGDE), 94.337–339
Entangled tree, 84.22
Entire functions, phase retrieval by,
  93.109–168
Entropy, 84.8; 91.37

Boltzmann entropy, 90.125
hybrid entropy, fuzzy sets, 88.252–256
information approach, 90.186, 189
Kullback–Leibler cross-entropy, 90.144,
  145
sample set, 84.264
Shannon's entropy, 90.125; 91.37, 38, 96
unified (r, s)-entropy, 91.41–42
  bivariate, 91.95–96, 98–102
  conditional, 91.95–107, 121–123
  multivariate, 91.95, 102–107
  properties, 91.73–75
Entropy coding, 82.117; 97.193–194
Envelope, signal description, 94.318
Envelope function, electron off-axis
  holography, 7.135; 12.31, 60;
  89.25–27, 29
Environmental chambers:
  at high voltage, 5.305, 320, 331, 335,
    338
  at raised temperature, 5.306
  dynamic, 5.320, 321, 322, 324, 329,
    331, 339, 341
  halo effect, 5.329
  inert gas medium, 5.328
  medium, 5.328–334
  for organic material, 5.299, 319–341
  oxygen effect, 5.328
  problems, 5.298
  protection, 5.328
  specimen:
    dehydration, 5.329, 331
    freezing, 5.333, 334, 335, 343
    homogeneity, 5.339
    hydration, 5.329, 331, 332, 333, 339,
      343
    motility, 5.339, 340
    support, 5.324, 325, 327
    thickness, 5.338, 339
    viability, 5.335–338
  static, 5.304, 320, 321, 323, 324, 325,
    326, 327, 336, 339
  temperature and, 5.333–335
  water, presence of, 5.329–333
  windows, 5.324

Environmental scanning electron microscope (ESEM), development, 96.510
Enzymes, 5.317
  activity reduction at interfaces, 8.113
  cytochemistry, 6.175, 204, 206, 208, 212
  distribution, 6.172, 173, 175, 176
    at myoneural junction, 6.204, 206
    quantitative assessment, 6.176, 177
  histochemistry, 6.171, 172, 216
    aims, 6.176, 177
    applications, 6.176
    development, 6.219–222
    with electron microscopy, 6.172, 173, 174, 176, 177, 178, 179, 181, 182, 183, 184, 186, 188, 191, 195, 196, 197, 198, 200, 202, 203, 204, 205, 209, 210, 214, 219, 220
    instrumentation and, 6.221, 222
    principles, 6.172
    specificity, 6.221
    technique, 6.220, 221
    with X-ray microanalysis, 6.221, 222
  localization, 2.332; 6.171
  radiation damage, 5.317, 318
  species variation, 6.178
Enzyme techniques:
  for acid phosphatase, 6.180, 181, 182, 183
  aldehyde fixation, 6.177, 178
  for alkaline phosphatase, 6.184–188
  capture reaction, 6.174
  control methods, 6.179
  dehydration, 6.177
  diffusion, 6.174, 175, 176, 178, 179
  dissection of tissue, 6.177
  embedding, 6.177
  fixation, 6.175, 177, 178
  FRP stability, 6.179
  incubation, 6.174, 175, 177, 178, 179
  localization, 6.174–176, 179
  methodology, 6.177–179
  nature of FRP, 6.176
  nomenclature, 6.174
  PRP release rate, 6.175, 178
  sectioning, 6.179
  for sulfatase, 6.181, 182, 183
  theory, 6.174–176
  tissue preparation, 6.177
  ultrastructure preservation, 6.177, 178
Enzyme tracer techniques, 6.217–219
EPI *See* Principle of extreme physical information
Epidotite, granoblastic texture, 7.37
Epi-illumination, 6.36
  gas laser source, 6.28
  with mirror objective, 6.28
  Ploem system, 6.28, 29
Epi-technoscope, 5.10
Epithelial cells:
  KFA phase contrast photomicrograph, 6.72, 73, 74, 77
  KFS phase contrast photomicrograph, 6.80, 81
  signal filtering, 8.18
Epithelial fibroblasts, SiO replication, 8.85
Epival microscope, 6.105
Epoxy resins, 2.322
EPPM *See* Extrapolated method of parallel projection
EPREVIEW (software), 13.26
EPR experiment *See* Einstein–Podolsky–Rosen (EPR) experiment
E-PROMs *See* Electrically altered ROMs
Equation for the reference structure, 90.252
Equilibrium contact angle studies, 8.97
Equivalence, fuzzy relations, 89.278–279
Equivalent eigenvectors, 90.65–67
Equivalent pass band, 7.139
Equivalent quantizers, 84.2
Erbium-doped fiber amplifiers (EDFAs), 99.71
Erosion:
  function, 88.203; 99.8
  mathematical morphology, 89.328–329, 343–344, 374
  multiscale dilation–erosion scale-space, 99.16–22
  set, 88.202

Erosion, *continued*
  support-limited, 88.219, 231
Erythrocytes, 5.45, 46, 342; 11.164
  catalase, three-dimensional structure, 7.341
  flow velocity, 6.13, 27
  hemoglobin, 6.162
  human, 6.160
  SiO replication, 8.85–86
Escherichia coli, 5.310, 314, 336
  stereo display, 7.10
  tracking microscopy, 7.9
ESEM *See* Environmental scanning electron microscope
ESETUP (software), 13.26
  equations for, 13.31–38
ESOLVE (software), 13.26, 31–32
  solving equations in, 13.38–39
ESRQs *See* Exhaustive search residual quantizers
Esterases:
  azo dye techniques, 6.187
  coupling azo dye technique, 6.197
  electron microscopy, 6.187, 188, 195, 196, 197, 198, 206
  functions, 6.195
  gold technique, 6.199
  histochemistry, 6.195, 196–199
  indoxyl technique, 6.197
  localization, 6.198
  nonspecific, 6.195, 197
    acetoxyquinoline/bismuth technique, 6.199
    indoxyl technique, 6.198
  optical microscopy, 6.195, 197
  specificity, 6.195
  substrates, 6.195, 196, 197
  thiocholine technique, 6.206
  thiolacetic acid technique, 6.198, 199
Estimation:
  Kalman filtering, 85.20–21, 56, 65
  probability law-estimation procedure, 90.145–146
ESVQs *See* Exhaustive search vector quantizers

Etching, vacuum microelectronics, 6.276; 83.43, 45, 46
Ethidium bromide:
  metachromasy of fluorescence, 8.53, 125
  nucleic acids and, 8.125
ETRAJ (software), 13.26
  charging effects on insulating specimens, 13.78
  direct electron ray tracing, 13.39–41
Euler–Cauchy formula, 13.202
Euler–Lagrange equation, 90.127
Europe:
  electron microscope production:
    chronology of microscope manufacture, 96.538–571
    manufacturers and model numbers, 96.535–538
    production and use trends, 96.571–573
  European Regional Meetings, 96.386–391
Eutectic ion sources, 11.113
Evanescent field, 12.281
Evanescent waves, 12.246, 283
Evaporated films, structure of, 3.178–183
Evaporators, 11.76
Event times, 90.16–23
Everhart–Thornley detector, 13.127; 83.223, 226
Evidence theory, and rough set theory, 94.171–182
Evolution, minimax algebra, 90.94–98, 100–102
Ewald's solution, 90.236
EWPP *See* Extremal-weight path problem
Exact identification filters, 94.384
Exchange energy, 82.199, 208, 210
Exchange principle, 89.277, 278
Exchange switch system, 91.201–204
Excitation error, 86.178, 180, 197
Excitation spectra:
  bandgap narrowing, 82.202
  fluorescence microspectroscopy, 14.132–134
Excluded middle, weakened law of, 89.260

Exclusion principle, electron optical coherence and, 7.172
Exclusive OR gate, 89.232–233
Exhaustive search residual quantizers (ESRQs), 84.5, 52–57
Exhaustive search vector quantizers (ESVQs), 84.3, 5, 22, 39–45
Expanded contrast technique, 6.297
Expansion function, 82.47, 50, 53, 54, 56, 58, 60, 62, 63, 66, 67
Expectation–maximization (EM) approach, image processing, 97.127–129
Expert systems:
 defined, 86.81
 for image processing and analysis, 86.124, 148; 87.87–114
Expert vision, 86.85–86, 164
Expitaxial thin films, 11.60
EXPLAIN (software), 86.95, 98, 100
Explicit learning, 94.286
Explicit representation, 84.180
Exponential filter, phase retrieval, 93.111–112, 116–118
Exponential map, 84.170, 84.185 et seq.
Exponential tails, 82.232, 244
Exposure:
 photomicrography:
  determination, 4.396–402
  long, 4.389–391, 400
 tools, 83.110
Exposure time, 1.192, 201; 7.87
Extended boundary, 84.219
Extended Cooley–Tukey fast Fourier transform, 93.47–52
Extended-GNC (E-GNC) algorithm, 97.132–136, 171–175
Extended Prony method, 94.369
Extended sources, transfer functions, 7.123
Extended two-particle model, in Coulomb interaction, 13.218
Extended unit cell, 86.191, 204, 205, 210
Extension function, 90.360
Extensive τ-mapping, 89.350–351
Extinction contours, 2.186, 233, 238
Extinction distance, for electrons, 2.183, 187
Extinction factor, 1.81, 82
Extinction point, 1.79, 85
Extinction thickness, 89.41
Extrapolated method of parallel approximate projections (EMOPAP):
 algorithm, 95.225–226
 convergence results, 95.226
 image recovery projection, 95.223–226
 problem statement, 95.223–224
Extrapolated method of parallel nonexpansive operators (EMOPNO):
 algorithm, 95.230
 convergence results, 95.230–231
 image recovery projection, 95.229–231
 problem statement, 95.229–230
Extrapolated method of parallel projections (EMOPP):
 control strategies, 95.218–219
 convergence results, 95.219–223
 image recovery projection, 95.213–223
 image restoration with bounded noise, 95.248
 iteration, 95.217
Extrapolated method of parallel subgradient projections (EMOPSP):
 algorithm, 95.228, 257
 control, 95.234
 convergence results, 95.229, 234, 256
 image recovery projection, 95.226–229, 252–253
 practical considerations, 95.234–235
 problem statement, 95.226–227
 relaxations, 95.235
 set theoretic formulation, 95.253–256
 stopping rule, 95.235
 subgradient projections, 95.228
 superiority to POCS, 95.234, 257
 weights, 95.234
Extrapolation, structural properties method, 94.391–392

Extremal-weight path problem (EWPP), 90.41–42, 45–46
Extrema product forms, 90.92–93
Extreme information, principle of, 90.139–147
Extremum sharpening filter, image enhancement, 92.32–34, 44–48
EZW algorithm, 97.221, 232

F

Fabrication techniques, 83.36
Fabry–Perot interferometer, 6.115; 8.34
Fabry–Perot resonance condition, 89.125, 126, 128
Face, finite topology, 84.201, 204; 86.228, 229
Facsimile machine, history, 91.204
Factorization, 85.23, 33–41, 63
Fading, imaging plate system, 99.258–259
Failure rule, LLVE, 86.121, 124
Fairchild Corp., semiconductor history, 91.145, 149, 150
Faraday's law, 82.25, 42
Far field radiation, 12.245
Far-out-of-focus holography, 98.373–387, 422–423
Fast atom bombardment, 11.76
Fast computation algorithm:
 of ICT, 88.57–58
 of Walsh transforms, 88.25
Fast convolution, 94.17–19
Fast Fourier transform (FFT) algorithm, 9.57–58; 84.62; 86.191, 202, 205, 210; 93.16
 Cooley–Tukey algorithm, 93.17–19, 27–30
  extended, 93.47–52
 Good–Thomas algorithm, 93.19–21
 reduced transform algorithm, 93.16–17, 21–27
Fast local labeling algorithm, 90.407, 410–412, 507

Fatty acid synthetase, three-dimensional reconstruction, 7.332
Fault lines, in conductive thin films, 83.33
FBLP *See* Forward–backward linear prediction (FBLP) method
FDM *See* Finite-difference methods
Feature frequency matrix, texture representation, 95.388, 390
 classification scheme, 95.402–403, 405
 distance measure, 95.402, 406
 feature image, 95.390–391
 generalized, 95.393
 moment feature vectors, 95.393, 400, 406
 one-dimensional, 95.393
 partitioned feature frequency vector, 95.400–401, 406
 two-dimensional, 95.393
FEE *See* Field emission
Feedback networks, 94.265, 276–278
Feedforward networks, 94.265, 276–278, 288
FEF *See* Function elimination filters
FEG *See* Field emission guns
Feigenbaum number, 94.282
Fejér-monotone sequence, 95.170–171
Feldspar, microtransforms, 7.49
FEM *See* Finite-element methods
Femtosecond-pulse measurement, phase retrieval, 93.167–168
Feret's diameter, 3.40, 46
Fermi–Dirac distribution, 83.18
Fermi energy, 82.209
Ferricyanide technique, electron microscopy, 6.214–216
Ferrite media, 5.149; 92.117, 132
 assembly process, 92.179–181
 chiral-ferrite media, 92.117, 119, 123, 126, 132
 circuit parameters, 92.181–183
 constitutive relations, 92.119, 120
 finite-element 2–D equations, 92.162–167
 Helmholtz wave equation, 92.159–162

Ferritin:
  structure, 3.200
  water layer thickness determination using, 8.81, 83–84
Ferroelectric domains, use of mirror electron microscopy, 4.226, 244
Ferroelectric liquid crystal, 91.248
Ferroelectric materials, mirror electron microscopy, 94.96
Ferromagnetic materials:
  electron microscopy, 7.190
  GS algorithm and, 7.224
  mirror electron microscopy, 94.97–98
Ferromagnetic microprobes:
  bulk probe properties, 87.138–148
  demagnetization coefficient, 87.141
  equipotentials, 87.142
  magnetic monopole and dipole moments, 87.146–147
  magnetocrystalline anisotropy field, 87.138–139
  o-contributions, 87.147–148
  radial stray field, 87.144–146
  shape-anisotropy field, 87.138–139
  vertical stray field, 87.142–144
  thin-film probes, 87.148–157
  equipotentials, 87.151–153
  magnetic moments, 87.155–156
  radial stray field, 87.154–157
  two-probe model, 87.149, 155
  type of sensors, 87.150
  vertical stray field, 87.152–154
Ferromagnets:
  interdomain boundaries, 87.176–180
  strong exchange interactions, 98.325
FET *See* Field-effect transistors
Feynman path integral method, 82.201, 239, 242
FFM *See* Feature frequency matrix
FFT *See* Fast Fourier transform (FFT) algorithm
Fiber optics, 2.12, 13; *See also* Optical fiber communications
Fiber plates:
  LLL-TV and, 9.51–52, 60
  radiation damage, 9.36, 51–52
  thermal treatment to eliminate opacity, 9.52
  TV image intensifier input stage, 9.34–36, 37
Fiber pulling, dye complexes, 8.62
Fibers, use of image analyzing computer, 4.381
Fibrillar proteins, electron microscopy, 7.282, 288
Fibroblastic animal cells, SiO replication, 8.85
Fibronectin, imaging, 10.318–319
Field-effect transistors (FET), 89.98, 136
  boundary conditions and, 91.216
  described, 91.149
  development, 91.150
  JFET, 86.32; 87.225, 227, 231, 245
  MESFET, 83.70, 71, 73, 75; 86.32, 36; 99.79
  MODFET, 86.54; 89.98
  MOSFET, 87.218, 226, 230, 245; 89.98; 91.150, 223
  QUADFET, 91.226–227
  vacuum fluorescent displays, 91.236
Field electron emission *See* Field emission
Field-electron microscope, 2.343, 344, 350
Field emission (FEE), 2.348; 83.3, 10, 12, 13, 16, 19, 33, 34, 35, 55, 67, 72, 87; 87.218
  adsorbate effects, 8.220
  analytical field calculation, 8.228–238
  angular emission distribution, 8.215–217
  band structure effects, 8.218
  brightness properties, 8.248–251
  cold field electron emission, 8.255; 9.191; 83.32
  conservation laws, 8.209
  crossover positions, 8.245
  current, 83.7, 55
  electron source, 83.3, 6, 222, 228, 255–257
  emission current density, 8.211, 212, 213
  field strength increase, adsorbate layers, 8.220

Field emission (FEE), *continued*
   fluctuations, 83.7, 58–61
   hot, 83.29, 30
   many-body effects, 8.218–220
   metal surface emission, 95.69–71,
      73–76, 78–81
      current density distribution, 95.74–76,
         78
      current stability, 95.78–81
      current-voltage characteristics,
         95.71–72
      electron potential energy, 95.65–66
      energy distribution of electrons,
         95.72–74
      extraction processes, 95.664365
      metal/vacuum barrier, 95.64–66
      thermionic emission, 95.66, 68–69,
         72–73, 75, 79
   nanotip *See* Nanotips
   numerical field calculation, 8.226–228
   point source model, 8.221–223
      applications, 8.255–256
   relativistic acceleration potential, 8.244
   spatial coherence measurement, 7.158
   sphere-on-orthogonal-cone (SOC)
      model, 8.223–225
   supply-function, 8.211–213
   system, 83.195
   temperature-field domains, 8.214
   theory, 8.208–220
      Fowler-Nordheim, 8.209
      modifications, 8.217–220
   thermionic emission in, 8.253
   tips *See* Microtips; Nanotips; Tips
   total current density, 8.211–213, 218,
      220
   total energy distribution, 8.214–215
   transmission probability, 8.210–211
   tunneling of electrons process, 8.208,
      209
Field-emission diodes, 8.238–241
   electric field, converging effect, electron
      beam, on, 8.238–240
   electrostatic lens, 8.239
   field strength, 8.238–239

   lens aberrations, 8.241
   meridional trajectory, electron field, of,
      8.239–240
   normal trajectory, electron beam, of,
      8.239–240
   spherical potential, 8.239
   virtual source, lateral extension of, 8.240
Field-emission guns, 8.139, 242–248;
      10.248; 11.46, 75; 13.164;
      95.63–64
   aberrations, 8.247–248, 257–258
   adsorption effects, work function, on,
      8.253
   applications, 8.208
   asymmetric electric lens, 8.242
   brightness properties, 8.248–251, 256
   cathode space electrostatic field
      equipotential lines, 8.247
   cathode stability, 8.252
   CD measurement with, 13.128
   crossover, 8.245, 251
   current density, 8.255–256
   destroyed cathode micrographs, 8.255
   dipole perturbation field, 8.230
   electrical breakdown, 8.255
   electron energy spread, 8.208
   electron optical properties, 8.221–251
   electrostatic anode lens optimization,
      8.242–243
   electrostatic triode schematic
      representation, 8.229
   emission current fluctuations, 8.253,
      255, 256
   emittance diagrams, 8.163–165
   energy exchange balance, 8.253
      protrusions, in, 8.255
   field calculation, 8.226–238
   first and second anode shapes, 8.242
   French development, 96.118
   holography application, 96.709–712
   Japanese contributions, 96.710–713
   magnetic focusing, 8.246–247
   mechanical stability, 8.252–253
   meridional section:
      cathode, through, 8.232–233

whole configuration, through, 8.229
    optimizing, 8.252, 258
    potential evaluation, 8.230
    probe current, 8.255
        theoretical design, 8.208
    scanning electron microscopes,
        96.143–144, 712–713
    sputtering, 8.253
    stability problems, 8.252–255
Field-emission microscopes, 2.344
    operation principles, 8.238
Field-emission tips, 93.58, 79
Field-emission triodes, 8.221
Field etching, 2.365
Field evaporation, 2.345, 357
    of impurities, 2.363
    rate of, 2.367
Field-free interval, 86.230
Field-induced barrier lowering, 86.27
Field ionization, 2.348, 351
Field-ion microscope, 2.343, 345, 391,
        397, 399; 8.238
    applications, 2.399
    bulk structure, 2.400
    computer analysis, 2.384, 390
    crystallographic analysis, 2.386
    gas supply, 2.353, 391
    image interpretation, 2.370 et seq.
    image recording, 2.393
    ion current, 2.355
    resolution of, 2.346
    specimen preparation, 2.399
    surface structure, 2.371, 401
    vacuum system, 2.396
Field ion sources, 11.110
Filamentary conduction, mirror electron
        microscopy, 94.129–131
Filar micrometer, 3.50
Films:
    grain, digital image processing,
        4.138–139
    indium–tin-oxide, 83.81
    Langmuir–Blodgett, 83.32
    and long exposure, 4.389–391
    molybdenum, 83.36, 37

moving, recording with, 7.75–79
    silver, 83.31
    thin, 83.28, 36, 49
Filtered backprojection (FBP), image
        reconstruction, 97.160–162
Filtered images, Fourier transforms and,
        7.19–29
Filtering:
    scale-space filtering, 99.2
    signals:
        convolution, 94.2–3, 4–8
        structural properties method,
            94.379–383, 385
Filtering algorithm, space-variant image
        restoration, 99.314
Filters:
    adaptive, 84.342–344
    adaptive extremum sharpening,
        92.44–48
    adaptive quantile, 92.14–16
    center-weight median, 92.17
    composite enhancement, 92.16–17
    echographic, image processing,
        84.342–344
    extremum sharpening, 92.32–34, 44–48
    iterative noise peak elimination,
        92.17–18
    linear, 92.55–62
        low-pass, 92.21
    majority, 92.65–68
    mathematical morphology, 89.351–354
    max/min-median, 92.40–42
    molecular, LVSEM images, 83.250
    morphological, 92.49–52, 75
    multistage one-dimensional, 92.39–44
    nonadaptive rank-selection, 92.66–68
    nonlinear mean, 92.18
    over-filters, 89.351–353
    polynomial, 92.68–70
    quadratic, 92.68
    rank-order, 92.14–18
    recursive, 92.21–22, 57
    soft morphological, 92.49–52
    two-dimensional, 88.326
    under-filters, 89.351, 353–354

Filters, *continued*
  zonal, 92.25
Fine magnetic particles, 98.408–409
  chromium dioxide, 98.415–422
  phase image simulations, 98.409–415
Fingerprints:
  database, image compression, 97.54
  morphological scale-space, 99.4–5, 29–30, 55
    equivalence, 99.30–32
    reduced, 99.32–37
  signal, 99.4–5
Finite abelian group, 93.3–6
  Fourier transform, 93.14–15
  vector space, 93.7–8
Finite-difference methods (FDM), 82.333–335
  basic formulation, 13.6–10
  comparison of, 13.171–173
  curved electrodes in, 13.8–9
  on electrostatic lens, 13.129, 153
  field calculation, 8.226–228, 243
  formulating equations in, 13.6–25
  Laplace's equation in, 13.7–8, 9–10
  magnetic formulation, 13.24–25
  on magnetic lens, 13.129, 153
  of modeling electron optical systems, 13.5–6
  multiple dielectric cuts, 13.21–23
  as numerical method, 13.155–157
  single dielectric cuts, 13.11–21
  software for, 13.31–32
  solving equations in, 13.38–39
  thermionic electron emission calculation, 8.228
Finite eigenvectors, 90.63–67
Finite-element mesh, 82.71, 77, 81, 87, 90, 92
Finite-element methods (FEM), 85.79
  anisotropic media, 92.132–134, 145–157, 183
  comparison of, 13.172–173
  computing methods in, 13.4–5
  on electrostatic lens, 13.129, 153
  field calculation, of, 8.226, 243
  on magnetic lens, 13.129
  of modeling electron optical systems, 13.2–3
  as numerical method, 13.157–162
  propagating structures, 92.94–95
  p-version, 82.361
  summary, 82.64
  2–D equations, ferrite media, 92.162–167
Finite energy spread, transfer function and, 7.129–134
Finite groups, 94.8–11, 55–57
Finite-impulse-response (FIR) filter, 94.4
Finiteness, minimax algebra, 90.14–16, 26–27
Finite sample size:
  effective, 84.271
  estimation, 84.262–263, 265–266, 289, 291, 296
  Fisher information matrix, 84.285–286
  initial conditions, importance, 84.298
  Morgera–Cooper coefficient, 84.271
  performance:
    adaptive pattern classification, 84.265–266, 270–271
    autoregressive parameter estimation, 84.292–293, 302–309
    covariance estimation, 84.287
Finite scalars, 90.5
Finite source size, transfer functions and, 7.124–129
Finite state scalar quantization, 97.203
Finite topology, image analysis and, 84.197–258
FIR filter *See* Finite-impulse-response filter
First fundamental form, 84.189
First fundamental form coefficients, 84.173
First-order perturbation, 90.253–254
Fisher information, 84.285–286; 90.124–125, 128–139
  additivity, 90.190–191
  classical electrodynamics, 90.149–153, 186–187, 196–198
  Cramèr–Rao inequality, 90.129–133, 165–166, 177–178, 187

general relativity, 90.170–174, 188
information divergence, 90.144–145,
    194–196
maximal information and minimal error,
    90.191–193
measure of disorder, 90.136–137
multidimensional parameters,
    90.133–135
new interpretation, 97.409
parameter estimation channel,
    90.128–129
Poisson information equation, 90.139
power spectral $1/f$ noise, 90.174–185,
    188
quantum mechanics, 90.154–165, 184,
    185, 187
scalar information, 90.135
shift-invariant case, 90.135–136
special relativity, 90.147–149, 186
uncertainty principle, 90.165–169,
    187–188
unified (r, s)-divergence, 91.41,
    115–120
    Csiszár's phi-divergence, 91.116–117
zero information, 90.143–144, 164, 185
Fission, 5.2
Fixation, 2.251, 293; 3.106, 141, 150;
    83.232–233
Fixation artifacts, 2.254
Fixatives, 2.255; 3.106
    aldehydes, 2.264, 286, 303, 326, 334
    osmium tetroxide, 2.255, 276, 283, 293
    potassium permanganate, 2.262, 287,
      310
Fixed image scanning, setting on a line by,
    8.31–35
Fixed point, 93.9
Fixed segment extraction technique,
    82.103
Flagella, tubulin structure, 7.341
Flash electron microscopy, 5.340, 343
"Flash" E-PROMS, 86.71
Flatband theory, 89.132
Flat gray-scale mapping, 89.367–368
Flat panel technology, 91.244–252

active matrix addressed display,
    91.248–251
electrochromic display, 91.251–252
electrophoretic display, 91.252
light controller display, 91.246–257
light emitter, 91.245–246
liquid crystal, 91.246–252
polymer dispersed liquid crystal film,
    91.248
subtractive display, 91.246–247
Flat screen, 91.236
Flat structuring function, 99.44
Flexibly coupled multiprocessor
    architectures, 87.290–291
Flicker noise, avalanche photodiodes
    (APDs), 99.74, 151–153, 154
Flight navigation, Kalman filtering and,
    85.8, 10, 54, 55, 58, 60–61, 67
Floating gate memory, 86.49, 71
Flow orientation, dye complexes,
    8.61–63
Flow velocity:
    moving graticule technique, 1.8
    moving grating technique, 1.28
    moving spot technique, 1.9, 31
    slit image technique, 1.29
    streak image technique, 1.26, 32
Floyd–Warshall algorithm, 90.54–56, 64
Fluctuations, 82.234; 83.58, 61
Fluid bubbles, carbon dioxide in water,
    98.44–61
Fluid dynamics, 85.79
Fluorescence, 14.122–127
    analysis with, advantages, 14.126–127
    in cytotoxic agents, 14.172–174
    defined, 14.122–124
    history, 14.125–126
    illumination, 6.21
    lifetime imaging, 14.199–204
    metachromasy, 8.53
    new methods in, 14.174–204
        confocal Raman microspectroscopy,
          14.198–199
        delayed fluorescence imaging,
          14.191–196

Fluorescence, *continued*
    fluorescence lifetime imaging,
        14.199–204
    multiwave three-dimensional
        imaging, 14.184–189
    phosphorescence, 14.191–196
    photobleaching recovery, 14.174–177
    polarized photobleaching recovery,
        14.180
    quantitative imaging, 14.183–184
    ratio imaging, 14.180–182
    relaxation microscopy, 14.196–198
    resonance energy transfer,
        14.189–191
    total reflection, 14.178–180
  ratio imaging, 14.180–182
Fluorescence microscopy, 6.18, 24, 117;
    8.54; 10.55–58
Fluorescence microspectroscopy,
    14.127–134
  applications, 14.134–174
    bioluminescent probes, 14.168–171
    cell cations, probing of, 14.160–167
    cell cytoskeleton, probing of,
        14.150–152
    cell metabolism, noninvasive probing,
        14.134–137
    cell organelles, probing of,
        14.137–160
    DNA, hybridization of, 14.172
    and electron microscopy, 14.172–174
    gene expression, 14.171–172
    probing methods, latest, 14.168–174
  confocal Raman microspectroscopy,
    14.196–199
  emission spectra, 14.127–132
  excitation spectra, 14.132–134
Fluorescent screens, 9.212–213
  with fiber plates, 9.34–36, 37, 58
  TV image intensifier input stage, 9.31,
    33–38
Fluorogenic probes:
  carbocyanines, 14.152
  of cell cytoskeleton, 14.150–152
  of endoplasmic reticulum, 14.147–150
  for Golgi apparatus, 14.140–141
  of lysosomal enzymes, 14.137–140
  of membrane potential, 14.152–158
  merocyanines, 14.158
  mitochondria, 14.141–147
  oxonols, 14.152–156
  proton indicators, 14.158–160
  styryl dyes, 14.156–158
Flux, electromagnetism, 82.8, 15, 16, 17
Fluxon, physical manifestations,
    5.282–285
Fluxon criterion, 5.270–272, 275
Fluxon unit, 5.268, 282
Flying spot microscopy, 1.30; 2.74, 80, 82,
    85, 88; 4.362; 10.2, 61–62, 70
Focal elements, 94.171, 173–180
Focus of attention, 86.128, 132
Focusing:
  cardinal elements of, 13.185–186
  charged particle, 89.392, 433–441
  echographic image process, 84.331, 344
  electron microscopy, error in, 1.125, 132
  first-order, in electron trajectories,
    13.181–186
  principal and fundamental trajectories,
    13.182–185
  scanning electron microscopy,
    13.113–119
Foldy–Wouthuysen representation, Dirac
    equation, 97.267–269, 322,
    341–347
Formaldehyde, fixation with, 2.265, 276
FORTRAN, 5.45; 85.273–275
Forward adaptation, in linear prediction,
    82.129
Forward–backward linear prediction
    (FBLP) method, 82.130; 94.370
Forward–backward loglikelihood, 84.296,
    298–302, 313–314
Forward difference formula, 93.253
Forward equations, 90.4–5
Forward Kalman filter, 85.49–50, 54, 69
Forward orbit, 90.13
Forward predictor, 84.299
Forward recursion, 90.4–9, 35

Forwardscattering, 86.179, 180, 193
Foucault contrast, 5.249
Foucault mode, Lorentz microscopy, 98.360–362
Four-body problem, melting points, 98.4–13
Four-electrode mirrors, energy analyzers, 89.421–425
Fourier analysis, 3.190; 88.6
Fourier–Bessel series, 13.135–136, 139
Fourier coefficients, 87.171
Fourier–Mellin transform, 84.159
Fourier modulus, phase retrieval, 93.110–112, 131–132, 144, 167
Fourier series expansion, phase retrieval, 93.118–124, 131–133, 140–142
Fourier space:
   in image processing with SEMs, 13.295–296
   three-dimensional reconstruction, 7.304, 310–314
Fourier spectrum, 12.27
Fourier's theorem, optical transmission and, 7.18
Fourier transform, 8.2; 88.6
   cross-grating produced, 8.16
   discrete, 84.62, 84
   dust contaminated wire mesh produced, 8.16–17
   electron microscopy, 3.169, 172, 215; 5.163, 164, 165, 167, 170, 172, 173, 184, 191, 197, 244, 271, 284, 286, 287, 288
   formation, 8.3–4
   group invariant algorithms:
      fast Fourier transform algorithm, 93.16–30, 47–52
      finite abelian group, 93.3, 14–15
      one-dimensional symmetry, 93.53–55
      three-dimensional symmetry, 93.1–53
   holograms, 10.140
      off-axis hologram, 12.26; 94.215–218, 244–246
   invariance, 84.135
   maxima, 8.15

optical diffraction analysis, 7.18–29
oval stacks, 7.28
pairs, 7.24, 88.172 et seq.
phase retrieval:
   blind-deconvolution problem, 93.146–148
   coherent imaging through turbulence, 93.155, 160
   Hartley transform and, 93.140
   stellar speckle interferometry, 93.143–144
   two-dimensional, 93.131
photochromic glass in the plane of, 8.18–19
point light source produced, 8.15
production, 7.19
production flow charts, 7.21
short-time Fourier transforms, 94.320
signal description, 94.320
truncated functions, 7.25
Fourth-order Hamiltonian function, power-series expansion, 91.8, 13
4,2 formula, 93.234, 237–239, 252, 253, 257, 261–262, 266, 277
Fowler–Nordheim equation, 2.349; 8.209, 218; 83.10, 12, 19, 47, 50, 55, 64, 91; 95.69–72, 99, 111–112
FPM *See* Fresnel projection microscopy
Fractal dimension, 88.204; 97.69–70
Fractal set, 88.204, 205
Fractal signal, 88.213
   fractal interpolation function, 88.214
   fractional Brownian motion, 88.216
   Weierstrass cosine function, 88.213
Fractures, quantitative particle modeling, 98.21, 22–30, 31–33
Frame buffer, 85.170, 176, 195
Frame of discernment, 94.171
Frame grabber, 85.170
Frame image feature, 86.117–119, 154
Frame store, 12.214; 91.242, 244
France:
   electron microscopy:
      biology, 96.93–100
      commercial development, 96.484

France, *continued*
   early pioneers, 96.93–95, 97–98, 101–102, 104–106
   electron guns, 96.118
   facilities, 96.96
   first application for financial support, 96.102–104
   high-voltage electron microscopy, 96.112–115, 117–120
   metallurgy, 96.106–110
   radiation defects, 96.118–119
   specimen holders, 96.117–118
   thin films, 96.110–112, 120
French Society for Electron Microscopy:
   meetings, 96.128
   origin, 96.128
   publications, 96.128–129, 861–876
   high voltage electron microscopy, 10.254–256
   historical contributions, electron probe microanalysis, 96.120–127
Françon–Nomarski variable phase contrast system, 6.104
Fraunhofer approximation, 94.202
Fraunhofer diffraction, 7.20; 10.140; 84.319
   and Abbe's theory of image formation, 13.271
   for circular aperture, 13.261–268
   electron wave optics, 13.261–268
   from discontinuous film, 5.257, 258
   in image transfer, 5.287, 289
   for isolated objects, 5.279
   in paraxial condition, 13.257–258
   for periodic magnetic structures, 5.256, 257
   for periodic structures, 5.284, 285
   for rectangular aperture, 13.258–261
   rotationally symmetric third-order aberrations and, 13.280
   theory, 5.261–263
Fraunhofer holograms, 7.166; 94.216, 227
Fredholm integral equation, 86.258, 260
Free Bloch waves, 90.296

Free propagation:
   charged-particle beam:
      scalar theory, 97.279–282
      spinor theory, 97.330–332
Free space Green's function, 90.334
Free will, 94.264
Freeze-drying, 2.252; 6.237, 238, 268; 8.114, 115–116, 122–124
Freeze-etching, 8.114, 122
Freeze-fracture, 8.123; 83.237–238, 242, 245, 253
Freeze-substitution, 2.252
Freezing-point depression, 2.291
Freezing techniques, 83.252; 96.740–741, 783–784
Frequency compounding, 84.342
Frequency domain speech coding, 82.158
Frequency-modulated (FM) signal, A-Psi GDE, 94.341
Fresnel approximation, 94.202, 203
Fresnel biprism, 12.53
Fresnel diffraction, 10.139–140; 12.39
   from domain walls, 5.282, 283
   and Helmholtz inequality, 5.272
   for isolated objects, 5.272–279
   in paraxial condition, 13.257–258
   for periodic structures, 5.284
   for rectangular aperture, 13.258–261
   TEM off-axis holography, 94.253–256
   theory, 5.261–263
Fresnel fringes, 7.158
Fresnel holography, 7.166; 12.53; 99.173
Fresnel–Kirchhoff formula, 13.254; 95.127, 128
Fresnel–Kirchhoff integral, 5.260, 261
Fresnel mode, 98.335–337
   geometric optics, 98.337–341
   wave optics, 98.341–350
Fresnel projection microscopy:
   coherence, 95.139, 140
   experimental procedures, 95.129, 130
   field emission current, 95.126, 127
   Fraunhofer diffraction, 95.129
   Fresnel diffraction, 95.127–129
   instrumentation, 95.124–126

irradiation effects, 95.143, 144
magnetic stray field, 95.142, 143
magnification factor, 95.124, 125
nanotip application, 95.124–126
   nanometric carbon fibers, 95.130–132
   ribonucleic acid, 95.132, 134, 136, 138, 144
   synthetic polymers, 95.132, 134, 136–139
   resolution, 95.140
   sample preparation, 95.132
   virtual projection point, 95.126, 127
Fresnel propagation factor, 7.194
Fresnel zone plates, 10.126–130
Friedel's law, 12.70; 89.17; 94.209
Friend leukemia virus, fixation and staining, 6.237
Fringe contrast, electron off-axis holography, 89.32
Fringe-setting, photoelectric, 8.42
Fringing, straight-edge response, 10.24
Frog:
   blood, 5.158, 159
      microdensitometry, 6.157
      phase contrast/fluorescence microscopy, 6.118
   erythrocyte nucleus, 6.160, 161
   heart, screen microkymography, 7.97
Frustum, representation of, 13.29–30
Full potential mode, 90.245, 248
Functional, defined, 86.128–129; 99.41
Function elimination filters (FEF), generating differential equations, 94.381–382
Function implementation, 88.91
   color conversion, 88.97
   convolution, 88.98
   differencing, 88.94
   efficiency, 88.104
   neighborhood operation, 88.101
   portability, 88.105
   Sobel operator, 88.101
   thresholding, 88.96
Function interface design, 88.74
Fundamental eigenvectors, 90.63–64

Fundamental theorem of surface theory, 84.173, 192
Fuzzy geometry, 88.256–260
   breadth, 88.258
   center of gravity, 88.259
   compactness, 88.257
   degree of adjacency, 88.260
   density, 88.259
   height, 88.257
   index, area coverage, 88.258
   length, 88.258
   major axis, 88.258–259
   minor axis, 88.259
   width, 88.257
Fuzzy inference engine, 89.317–318
Fuzzy knowledge base, 89.313–316
Fuzzy relational calculus, 89.266, 276–289
Fuzzy relations:
   binary relations, 89.292–293
   calculus, 89.266, 276–289
   closures and interiors, 89.293–294
   likeness relations, 89.296–297
   similarity relations, 89.294–296
   theory, 89.255–264
Fuzzy set theory, 89.255–264
   applications, 89.264–322
   image analysis, 88.247–291
   rough sets and, 94.151

G

Gabor, Dennis, 12.31; 91.259–283; 96.139, 398
Gabor–DCT transform, 97.52
Gabor expansion:
   biorthogonal functions, 97.27–29
   exact Gabor expansion, 97.23–30
   image enhancement, 97.54–55
   quasicomplete, 97.34–37
Gabor focus, 12.49
Gabor functions (Gabor wavelets, Gaussian wave packets, GW), 97.3–4, 5, 7
   applications, 97.78

Gabor functions (Gabor wavelets,
   Gaussian wave packets, GW),
   continued
   human visual system modeling,
      97.41–45
   continuous signal, 97.23–27
      biorthogonal functions, 97.27–29
      Zak transform, 97.29–30
   discrete signals, 97.30–33
      Daugman's neural network, 97.31–32
      direct method, 97.32–33
   drawbacks, 97.6–7
   image analysis and machine vision,
      97.61–78
   image coding, 97.50–54
   image enhancement, 97.54–59
   image reconstruction, 97.59–60
   machine vision, 97.61–66
   mathematical expression, 97.5
   orthogonality, 97.6, 11, 13, 22
   quasicomplete Gabor transform,
      97.34–37
   receptive field of visual cortical cells,
      97.41–45
   vision modeling, 97.17, 34–35, 41–45
Gabor hologram, 10.139, 140–144
Gabor transform, quasicomplete, 97.34–37
Gain, defined, 99.76
Gain–bandwidth product (GBW), 99.86,
   110
Galena, 6.290
Galerkin's equations, 82.47, 49, 50, 52, 57
Galerkin's method:
   application to potential formulations,
      82.49
   in eddy current case, 82.53
   general description of, 82.46
   in static case, 82.50
   for waveguide and cavity, 82.61
Gallium arsenide:
   DRAM cells, 86.4–74
   field-effect transistor (GaAsFET),
      87.227, 244; 91.216
   integrated circuit technology, 86.1–4
   properties, 92.82, 117

semicondutor history, 91.175–184
Gallium phosphide, 91.179
Galvanomagnetic devices, 91.174–175
Gamma (minimax algebra), 90.46–47
Gamma (photographic), 1.187
Gamow factor, 8.210
Gases:
   adsorption and desorption, 83.63
   refractive index gradient, 8.11
Gas evaporation method, Japanese
   contributions, 96.785–786
Gas inlet system, 11.91
Gas laser, 91.179
Gas stirring, 1.60
Gate, 83.30, 43, 47, 49, 78, 80, 85
   disruption, 83.63, 65
   leakage, 86.37
   logic, spin-polarized single electrons,
      89.229–232
Gauge transformation, 82.17
Gaussian algorithm, 85.37, 43
Gaussian beam, 93.184–185, 186
Gaussian curvature, 84.174, 191
Gaussian derivatives:
   edge detection, 97.64
   vision modeling, 97.17, 19, 45
Gaussian distribution, in Coulomb
   interaction, 13.219, 221
Gaussian elimination:
   in direct electron ray tracing, 13.41
   in FDM methods, 13.10
Gaussian filter, 99.2–3
Gaussian fit, 86.221
Gaussian Markov random fields (GMRFs),
   97.107
Gaussian process, 85.17, 19
Gaussian scale-space, 99.3–5, 55
Gaussian statistics, 5.69; 82.234
Gaussian trajectory, in Hamiltonian
   representation, 91.13–15, 30
Gaussian wavefront, propagation,
   93.204–207
Gaussian wavelets:
   machine vision, 97.61, 63
   texture analysis, 97.64–68

Gaussian wave packets *See* Gabor functions
Gauss's theorem, 82.24, 25, 40
Gauss–Weingarten equations, 84.176, 191
GBM *See* Generalized Boltzmann machine
GBW *See* Gain-bandwidth product
GDE *See* Generating differential equations
G.E.C. Plessy Semiconductors, 91.150, 168, 223
Gedanken experiment, 90.125, 133, 138, 147
   classical electrodynamics, 90.149–150
   general relativity, 90.170, 171
   power spectral $1/f$ noise, 90.176–177, 188
   quantum mechanics, 90.155, 165
   special relativity, 90.147
Gelatin, 5.317
Gene expression, fluorescent probes for, 14.171–172
General approximation, minimax algebra, 90.111–112
General basis algorithm, 89.337
   gray-scale function mapping:
      dual basis, 89.371–372
      opening and closing, 89.372–374
   τ-mapping:
      cascaded mappings, 89.345–348
      dilation, 89.343–345
      dual basis, 89.348–349
      erosion, 89.343–344
      intersection, 89.339–343
      translation, 89.337–338
      union, 89.338–339, 342–343
   translation-invariant mapping:
      cascaded mapping, 89.363–365
      dual basis, 89.363
General decomposition algorithm, 89.371
Generalized Boltzmann machine (GBM), 97.150–151
Generalized Chernoff measure, 91.125
Generalized cylinder, 86.154–155
Generalized expectation-maximization (GEM) algorithm:
   image processing, 97.93, 153
   tomographic reconstruction, 97.162–166
Generalized information measures, 91.37–41
   unified r, s-information measures, 91.41–132
Generalized integration transformation, eikonals electrostatic
   lenses, 97.396–403
   magnetic lenses, 97.369–381, 389–392
Generalized Lloyd algorithm, 84.10
Generalized Lloyd–Max conditions, 84.9
Generalized matrix, 90.112–113
Generalized matrix product, 84.67
Generalized phase planes (GPP):
   signal description, 94.324–325, 347–365
   signal detection, 94.379
Generalized probability density function, 91.112
Generalized quantizer, 94.386
Generalized sampling, 94.386
Generalized soft dilation, 92.52
Generalized soft erosion, 92.52
Generalized spectral theory, 94.319
Generalized transitive closure, 90.79–80
General linear dependence, minimax algebra, 90.114–117
General relativity, information approach, 90.170–174, 188
General theory of image formation, 93.174–176, 215–216
Generating differential equations (GDE):
   A-Psi type, 94.339–347, 388
   data compression, 94.387–391
   ensemble GDE, 94.337–339
   parameter estimation, 94.366
   signal description, 94.324, 327, 336–347
   three coeffecients, 94.343–347
Generating phase trajectories, 94.324
Generation lifetime, 86.13, 29
Generation-limited memory, 86.49, 71
Generation width, 86.15
Genetic algorithms, 88.288–290
Geologic maps, Pennsylvania, 7.41
Geometric algebra, 95.272; *See also* Spacetime algebra

Geometrical optics:
  differential phase contrast mode,
    98.351–357
  Fresnel mode, 98.337–341
Geometric correction, computer image
    processing, 4.89
Geometric relations, 86.143, 146–151, 164
Geometric theorem, 86.86, 149, 151, 153, 164
Geometric transformation, 85.182
Gerchberg–Papoulis algorithm, 87.14, 16,
    20; 95.178–179
Gerchberg–Saxton algorithm (G–S
    algorithm), 7.220, 224, 240, 243;
    13.297–298; 93.110
German Democratic Republic; See also
    Germany
  electron microscopy:
    biology, 96.175
    materials science, 96.175, 178
    metallurgy, 96.175
    microscope production, 96.171–172,
      175, 448–449
    international activity, 96.179–180
    monographs, 96.178–179
    national conferences, 96.178, 180
    resignation from German Society for
      Electron Microscopy, 96.157
    societies, 96.178
Germanium:
  amorphous films, lattice fringes, 7.156
  dislocations, mirror electron microscopy
    image, 4.247
  properties, 92.82
  semiconductor history, 91.143, 144, 149,
    171, 173
Germanium avalanche photodiodes,
    99.89–90
German Society for Electron Microscopy:
  board members:
    original members, 96.151, 153
    table, 96.165–168, 802–803
  constitution, 96.153, 155, 161–165
  dues, 96.158–159
  Ernst Ruska Prize Foundation, 96.158
  honorary members, 96.170, 803
  meetings:
    first conference, topics, 96.151–152
    international conferences, 96.155–157
    locations, 96.802
    minutes of first meeting, 96.151,
      159–160
    origins, 96.150–153, 155–159, 801
    publications, 96.157–158, 852–876
    working groups, 96.157
Germany; See also German Democratic
    Republic
  electron microscopy, 10.218–221;
    96.149–150
  electrostatic electron microscope,
    96.137–141
  field emission microscope,
    96.143–144
  magnetic electron microscope,
    96.132–137
  scanning electron microscope,
    96.142–143
  Knoll, Max, 83.204; 95.13–14, 16–18,
    32, 42; 96.132–135, 142
Gestalt law, 86.111 112
Giant magnetoresistance, magnetic
    multilayers, 98.400, 402
Gibbs distributions, Markov random fields,
    97.106–108
Gibbs sampler algorithm, 97.119–120, 123
Ginzberg–Landau equations, 87.172–173
Ginzberg–Landau theory, 14.9–11
Givens transformation, 85.28–32, 37, 43,
    53, 62
Glaser, Walter, 96.59–66
  contributions to electron microscopy,
    96.59–66, 795
  Grundlagen der Elektronenoptik,
    96.59–64, 66
Glaser model, axial flux density,
    13.143–144
Glaser–Schiske diffraction integral,
    93.195–202
Glass, optical fiber technology, 91.199–201
Glass spheres, Ronchi-grid
    photomicrographs, 8.13

Glassy carbon field emission cathodes, 8.252, 257
Global covering, 94.156, 185
   LERS LEM1 option, 94.188–190, 191
Global match, 86.156
Global positioning system (GPS), 85.54–55, 62, 66
Global predicate, 84.236–239, 245
Global processing plan, 86.95–96, 98, 105
Global reduce operation, 84.79; 90.361
Global warming, LERS, 94.194
Globular macromolecules, 8.113
Globular proteins:
   negative staining, 7.288
   three-dimensional structure, 7.282
   tubular crystals, three-dimensional structure, 7.339–340
GLOL, 5.44, 45, 62, 75, 76, 77
   coding, 5.65
   command structure, 5.88–93
   Karnaugh map, 5.80
   in measurement analysis, 5.69–72
   procedures, 5.66, 67, 68
   sets, 5.80
   shrinking, 5.87, 88
   simple statements, 5.80–88
   structure, 5.45
   uses, 5.63, 65
GLOPR computer, 5.68, 69
Glove box, 5.3
Glow discharge:
   apparatus, 8.130–132
   support film treatment by, 8.120, 127–130
Glucose oxidase, three-dimensional structure, 7.339, 341–344, 345
Glucose-6-phosphatase, histochemical technique, 6.184, 185, 186
Glutaralehyde, 2.268, 303
Glycogen, fixation, 2.262, 263, 270, 274, 280
GMRFs *See* Gaussian Markov random fields
GNC algorithm *See* Graduated nonconvexity algorithm
GODPA system, 85.161–162, 166, 175
Goguen implication operator, 89.280
Gohberg–Semencul decomposition, 84.296
Golay pattern transformation, in CELLSCAN/GLOPR system, 5.62–68
Gold:
   colloidal, imaging technique, 10.318–319
   colloidal gold labeling, SEM, 83.214–216, 226–227
   crystal-aperture scanning transmission electron microscopy, 93.59, 90–105
   energy selected images, 4.351–352, 353
   grid, electron probe microanalysis, 6.280, 286
Gold–palladium cracks, 83.182
Golgi apparatus, 6.181, 183, 189
   discovery, 96.97
   probes for, 14.140–141
Gomori method:
   for acid phosphatase, 6.180, 191
   for alkaline phosphatase, 6.172, 173, 174, 184, 191
Gondran and Minoux theorem, 90.115–117
Good–Thomas (GT) algorithm, 93.2
   fast Fourier transform algorithm, 93.19–21
   hybrid RT/GT algorithm, 93.25–26
Gopinath formula, 12.173
GPP *See* Generalized phase planes
Grad B drift, 12.124
Gradient, defined, 88.304
Gradient functions, homotopy modification, 99.46, 48–51, 54
Gradient-index optics, 8.6–7, 20
Gradient watershed region, scale-space, 99.51–53
Graduated nonconvexity (GNC) algorithm, image processing, 97.90, 91, 93, 124–127, 153, 168–175
Graham relation, 87.106
Grain boundaries, 2.231, 386; 7.52
Grains, optical diffraction analysis, 7.33
Grain shape, 3.35

INDEX

Grain size, photographic, 2.153; 7.35
Grains per electron in emulsions, 1.194, 197
Grain yield, autoradiography, 2.156
Grandmother neurons, 94.284
Granite, transforms, 7.47
Granular electron devices, 89.203–208, 244
  quantum-coupled, shortcomings, 89.221, 244–245
Granular electron transistors, 89.205–208
"Granular" flow, 1.3
Granularity:
  data, rough set theory, 94.151–194
  emulsions, 1.196
  imaging plate system, 99.259–262
  stains for biological macromolecules, 7.290
Granular materials, 87.210, 219, 231, 234, 239, 243
Granulite, 7.37
Granulocytes, 5.46; 6.77
Granulometries, 99.11
Graphical user interface (GUI), 88.107
Graphic display, 91.234–236, 237, 244
Graphite, 5.146
  crystalline, 5.308
  liquid drop formation on, 98.30, 32–44
Graphs:
  acyclic graphs, 90.52–54
  and-or graphs, 86.138
  connected graphs, 90.48–51
  directed graphs, 90.36–41
  theory, 84.107, 113
  underlying finite graph, 90.38
GRASP library, 94.63–78
Grassman algebra, quantum theory, 95.377–379
Graticules, 1.7, 8, 10; 3.41, 42, 44, 47, 73–75, 79–80
Gray levels:
  and final image, 10.10
  porosity analysis, 93.308–311
Gray level statistics, 84.327, 328, 337
  cooccurrence matrix, 84.338–339
  first-order, 84.329–330
  histogram, 84.328–329
  kurtosis, 84.329–330
  mean, 84.329–330
  Rician variance, 84.333, 336
  second-order, 84.328, 337
  signal-to-noise ratio, 84.329, 333
  standard deviation, 84.329–330
  structural variance, 84.336
Gray model, axial flux density, 13.141–142, 144
Gray-scale images, 91.242, 244, 252
  mathematical morphology, 99.39
Gray-scale morphology, 89.326, 336; 99.8
Gray-scale reconstruction, 99.48–49
Gray-scale τ-mapping, 89.336–337, 366–374
  general basis algorithm, 89.370–375
Gray-scale transformations, histogram equalization, 92.6–9
Gray value, 92.4–5, 9, 19, 20, 27, 38
Gray-value function, 92.6, 29, 62–63
Gray-value transformations, 92.4–9
Great Britain *See* United Kingdom
Greatest lower bound, 84.68
Greatest-weight path problem (GWPP), 90.41–42
  matrix power series, 90.80, 84
  max algebra, 90.42–44
Green's function, 82.210, 225, 242; 98.81–82
  dyadic:
    admittance, 92.108, 112
    chiral-ferrite media, 92.123–125
    impedance, 92.103–106, 108
  electron diffraction, 90.334–338
  Galerkin's method, 82.44, 45, 47–54, 56, 57
  microstrip circulators, 98.128
    three-dimensional model, 98.219–238, 316
    two-dimensional model, 98.79–81, 98–103, 121–127, 316
  nonrelativistic free particle, 97.280, 350–351

time-dependent quadratic Hamiltonian, 97.351–355
Grigson coils, 94.221
GRIN lenses, 8.6–7
Grivet–Lenz model, axial flux density, 13.142, 144
Grobner basis, 86.151–153
Ground–ground support information, 85.54–55, 62, 66
Ground state computing, 89.218–220
  antiferromagnetism, 89.226–228
Group algebra, 94.10
  convolution by sections, 94.4–8
  direct product group algebra, 94.14–19
  finite groups and index notation, 94.8–10
  GRASP class library, 94.55–63
    source code, 94.63–78
  group convolution, 94.3, 11–12
  history, 94.3
  inverses of group algebra elements, 94.22–27
  matrix representation, 94.12–14
  matrix transforms, 94.27–44
  semidirect product group algebra, 94.19–22
  signal and image processing, 94.44–54
Group algebra class, image processing, 94.58–60
Group algebra matrixes class, image processing, 94.60–62
Group class, image processing, 94.55–57
Group convolution, filtering signals, 94.3, 11–12
Group elements, matrix representation, 94.12–14, 24
Group-invariant transform algorithms, 93.1–55
Group theory:
  affine group, 93.10–11
  character group, 93.6–9
  electron optics, 85.231–256
  finite abelian group, 93.3–6
  point group, 93.9–10
Group III–V compounds, semiconductors, 91.171–188

G–S algorithm *See* Gerchberg–Saxton algorithm
GT algorithm *See* Good–Thomas algorithm
GT–RT algorithms, 93.45–46
GUI *See* Graphical user interface
Guided missiles, 91.149, 174
Guided-wave structures, 92.183–200
GWPP *See* Greatest-weight path problem
Gyro, 85.2–7
  mechanical gyro, 85.4–5
  optical gyro, 85.5–7
Gyroelectric-gyromagnetic variational analysis, 92.136–141

H

Hadamard matrix, 97.217–218
Hairpin tungsten filaments, 9.191–192
  grid cap design, brightness and, 9.192
Half-plane aperture images, phase problems and, 7.262
Hall, Cecil, contributions to electron microscopy, 96.81–82
Hall effect, 91.175
  anomalous, mesoscopic devices, 91.224
  quantized, 91.223–224
Halo effect, 6.58–62, 87, 88, 90, 98, 101, 104, 129
Halperin–Lax band tails, 82.234–239
Hamaker approach, 87.54–55
Hamaker constant, 87.72–87
  density-modulated, 87.117
  electric force microscopy, 87.132–133
  entropic, 87.58, 77, 81
  macroscopic, 87.93
  nonretarded, 87.59–60, 62, 77–78
    evaluation, 87.75–76
    as function of absorption frequencies, 87.78
    as function of optical refractive indexes, 87.76
    spectral contributions, 87.74–75
  oscillating, 87.116

Hamaker constant, *continued*
  partial van der Waals forces pressures, 87.91
  particle–substrate, 87.99
  retarded, 87.59, 63
    as function of effective refractive indexes, 87.79–80
    metallic half space, 87.64
    sensitivity to spectral features, 87.74
Hamilton analogy, 5.268
Hamiltonian equation, 91.4
Hamiltonian functions:
  power-series expansions, 91.8–13
  up to tenth-order approximations, 91.4–13
Hamiltonian mechanics, 91.4
Hamiltonian representation, Gaussian trajectory, 91.13–15, 30
Hamilton–Jacobi equation, 5.269
Hamming distance, 89.305
Hamming's modified predictor-corrector method, 13.203, 205, 207
Hamster chromosomes, SiO replication, 8.96, 98–99
Hankel matrix, 90.119
Harmonic detection, setting on a line by, 8.33–34
Harmonic microscopy, 10.84, 85–87
Hartley transform, 93.139–143
Hartree–Fock equation, 86.221
Hartree potential, 86.181, 221
Harvard architecture, 85.176
Hashimoto, Hatsujiro:
  career:
    Cambridge, 96.607–611
    Hiroshima, 96.598–599
    Kyoto, 96.599–604, 612–618
    Osaka, 96.619–620, 622–623
  electron microscopy history, 96.597, 605–607, 626–631
Hausdorff dimension, 88.204
HBT *See* Heterojunction bipolar transistor
H and D curve:
  for charged particles, 1.188
  for light, 1.185

HDTV *See* High-definition television
Health physics, 5.3, 5
Hearing, 85.81
Heart:
  beating, 6.2, 38, 39, 44–46
  contractions, stripe kymography and, 7.96
  microcirculation, 6.36–38
  ventricular muscle, 6.28
Heat equation, 99.6
Heavy atoms, phase information and, 7.270–272
Hebb's rule, 94.278, 281, 286
Height, fuzzy set theory, 89.265–266
Heine condenser, 6.116
Heisenberg inequality, 5.270
  in Fraunhofer diffraction, 5.270, 271, 272
  in Fresnel diffraction, 5.272
Heisenberg principle, information approach, 90.167, 169, 187–188
Heisenberg uncertainty, crystal-aperture scanning transmission electron microscopy, 93.57–58, 64
Helical symmetry, biological macromolecules, 7.209, 337–339
Helium:
  melting point, 98.11–13
  as ultraviolet light source, 10.271
Helium–neon light:
  interference image, 10.77
  optoelectronics, 10.66
  pulsed lasers, 10.197
Hellinger's distance, 91.43
Helmholtz equation, 5.268, 269
Helmholtz–Lagrange theorem, 13.185
Helmholtz theorem, 5.261; 13.176, 197
Helmholtz wave equation:
  ferrite media, 92.159–162
  three-dimensional circulator model, 98.137–138
  two-dimensional circulator model, 98.86–87
Hemagglutinin, electron microscopy, 7.258
Hematin, EELS, 9.75, 76

Heme proteins, DAB technique, 6.210, 211
Hemispherical detectors, 12.186
Hemispherical ES analyzer, 12.109
Hemocyanin, three-dimensional structure, 7.349–352
Hemoglobin, 6.211; 10.310
HEMT *See* High electron mobility transistor
Hermitian object functions, phase retrieval, 93.123–124, 126, 133
Herpes virus, three-dimensional reconstruction, 7.367, 368
Hessian matrix, 84.282, 303
Heterochromatin, higher-order structure, 8.95, 97, 100–101, 103
Heterocyclic dyes:
　absorption and emission characteristics, 8.58
　extinction coefficient, 8.58
　metachromasy of fluorescence, 8.53
　purification, 8.58
Heterodyne microscopy, 10.82–84
Heterogeneous algebra, 84.72
Heterojunction bipolar transistor (HBT), 99.79
Heterojunction memory, 86.44
Heun formula, 13.202
Heuristic approach, multislice approach to lens analysis, 93.173–216
Hexagonal-oriented quadrature pyramid, joint space-frequency representations, 97.19, 20
Hexamethylbenzene, electron density map, 4.115, 118
Hidden line, 85.114
Hidden surface, 85.114–198
Hidden volume, 85.114
Hierarchical bases, 82.351
Hierarchical bus architecture, 87.282–283
High-beam energy scan:
　charging effects on insulating specimens, 13.79–86
　collection efficiency, 13.82–85
　computation, potential, 13.79–82
　and secondary electron computation, 13.82
　simulated line scan, 13.85–86
High-definition television (HDTV), 83.6, 77, 79; 88.2
High-density limit, 82.200
High-density regime, 82.207, 232
High electron mobility transistor (HEMT), 91.186; 99.79
Higher-order spectra, signal description, 94.322–324
High-frequency lenses, 1.224
High-frequency scanning techniques, 8.36
High-order Laue zone (HOLZ), 11.47; 90.239–240, 302–303
　emulsion transfer response and, 9.215
High-resolution electron microscopy (HREM), 9.179–218; 10.233, 256–257; 11.3; 90.216
　amorphous film structure, 9.189–190
　astigmatism correction, 9.202–203
　atom clusters, image contrast theory and, 9.189–190
　beam current, energy spread and, 9.191
　beam-related damage, 9.209–211
　condenser aperture, 9.195–197
　contamination, 9.204–206
　contrast transfer function and, 9.186
　dark-field microscopy, 9.189
　developments, 13.244–245
　electron micrograph analysis, 9.214–218
　enhanced shadow contrast, objective aperture and, 9.204
　fluorescent screens and, 9.212–213
　focus fluctuations, 9.206–207
　hairpin tungsten filaments, 9.191–192
　heavy atom clusters and, 9.182, 202
　high-resolution images, 9.211–214
　holey films and, 9.182, 197, 198, 199, 205, 209–210
　illumination system:
　　adjustment, 9.190–197
　　brightness, 9.190
　　high-brightness illumination sources, 9.191

High-resolution electron microscopy (HREM), *continued*
   hollow cone illumination, 9.196–197
   tilted beam dark-field illumination, 9.196, 197
   image contrast theory, 9.183–190
   image intensifiers and, 9.211
   image plate with, 99.274–285
   imaging system alignment, 9.197–204
   Japanese contributions, 96.759–765
   lens aberrations, alignment and, 9.197
   low dosage, 88.127, 158
   many-beam crystal images, 9.186–189
   micrograph viewing conditions, 9.217–218
   objective apertures, 9.204
   object supports, 9.208
   operation, 9.180–182
   optical binoculars and, 9.211–212
   phase-shifting apertures, 9.204
   plane-wave illumination, condenser aperture and, 9.196
   pointed oriented single-crystal filaments, 9.192–195
   problems, 9.204–207
   specimen:
      electrostatic charging, 9.210–211
      limitations, 9.207–211
   specimen stage vibration/drift, 9.207
   spot size, illumination system, 9.190
   test objects, 9.182, 207–208, 209
   thin film apertures, 9.204
   through-focus series, 9.216–217
   tilting of crystal films, contamination and, 9.209
   two-beam crystal lattice fringes and, 9.183–185
   United States contributions, 96.367, 369
High-resolution profile imaging, 11.67
High-temperature superconductivity, 14.26–40
   applications, 14.39–40
   critical field and critical current density, 14.31–36
   electron pairing, mechanisms for, 14.38–39
   elements, substitution of, 14.30–31
   layered structure and anisotropy, 14.29–30
   magnetic shielding, 14.38
   materials for, 14.26–29
   potential applications, 14.113–116
   present state of, 14.110–116
   properties of, 14.29–38
High-voltage electron microscopy, 1.146; 2.167 et seq., 221; 5.305, 308, 309, 312, 318, 320; 10.251–252, 254–256
   advantages, 2.173, 201
   applications, 2.231
   chromatic aberration, 2.193
   commercial production in Europe, 96.440–441, 492–494
   contrast in, 2.204
   design, 2.213
   diffraction in, 2.197
   French contributions, 96.112–115, 117–120
   generators for, 2.213, 220
   history, 96.394–395
   Japanese contributions, 96.612–615, 700–704, 736–737, 754, 756–759
   microchamber, 2.242
   object, effect on, 2.196
   objective lens, 2.174, 224, 229
   resolving power, 2.173, 175, 194, 224
   United States contributions, 96.366, 369
Hilbert function, 93.115
Hilbert phase, 93.115–116
Hilbert space *See* Convex feasibility problem; Convex set theoretic image recovery
Hilbert transform, 7.230, 231, 237, 242, 243; 93.111–116, 120–121
Hilbert transform pair, 88.308
Hildebrandt–Schiske procedure, 13.145
Hilger and Watts microscopes, 8.30, 35
H-invariance, 93.10
Hiroshima University, 96.598–599
Histochemistry, 83.213–214
   of ATPases, 6.189–195

of cholinesterases, 6.200–203
with electron microscopy, 6.172–174, 176–179, 181–184, 186, 188, 191, 195–198, 200, 202–205, 209, 210, 214, 219, 220
of enzymes, 6.171, 172, 176, 177, 216, 219–222
of esterases, 6.187, 195–199
with optical microscopy, 6.172–175, 181, 184, 186, 191, 195, 201, 202, 209, 213
of oxidation-reduction enzymes, 6.209–216
of phosphatases, 6.179–188
of sulfatases, 6.181
techniques, 6.174–176, 177–179
Histogram equalization, 10.10; 92.6–9, 28–30
Histogram stretching, 92.27–28
Histological specific staining, 6.276
Hitachi:
　800X, 10.257
　HF-2000, 96.716
　HFS-2, 96.712
　HS series, 96.688
　HU-1, 96.653–654
　HU-2, 96.654–655
　research:
　　anticontamination devices, 96.697–698
　　electron energy loss, 96.691–692
　　pointed cathode, 96.690–691
　S-900, 96.713
　World War II impact, 96.655
Hit-or-miss mapping, 89.358
Hodgkin–Huxley neuronal model, 94.279–280
Holder inequality, 91.46, 94–95
Holder's inequalities, 91.46, 94–95
Hollow cone mode, 11.25
Hologram fringes, electron off-axis holography, 89.23–24, 32
Holograms, 12.36, 73, 86; 91.260, 276, 278, 280; 99.173–174; *See also* Holography

classification, 10.101, 139–140
defined, 89.8
Denisyuk hologram, 10.139
diffraction gratings, 10.117–126
empty holograms, 10.167
Fraunhofer holograms, 94.216, 227
Gabor hologram, 10.139
image plane holograms, 10.149, 162–163
image reconstruction, 99.200–205, 207–210, 218–221
　double-exposure electron hologram, 99.207–210
　in-line optical bench, 99.200–203, 207
　Mach–Zender interferometer, 99.203–215, 219
　optical symbolic substitution, 89.74, 77–79
　recording, electron holograms, 89.32–35, 47; 99.192–200, 207–210
　ultrasonic, 4.258
Holographic display, 91.254
Holographic microscopy, 10.208–209
　applications, 10.190–208
　　biomedical, 10.190–194
　　materials science, 10.194–196
　　particle analysis, 10.196–208
　CCTV system, 10.197
　characteristics and principles, 10.100–101, 104–139
　coherent noise elimination, 10.159–173
　　patterns, 10.159–161
　　techniques, 10.161–163
　　unidirectional suppression, 10.163–173
　collimating transfer system, 10.200–201
　equation of holography, 10.134–135
　fog droplets, 10.201–202
　Fourier transform, 10.103
　Fresnel zone plates, 10.126–130
　Gabor in-line system, 10.140–144
　historical note, 10.101–104
　image reconstruction, 10.133–139
　in-line microscopy, 10.140–144

Holographic microscopy, *continued*
  lasers, types, 10.197, 204
  lateral or transverse magnification, general formula, 10.141
  lensless systems, 10.140–148
  Mach–Zender type, 10.173, 176
  MGI-1, 10.153–154, 194, 195
  Michelson interferometer, 10.176
  microholographic systems, direct wavefront reconstruction, 10.148–153
    in-line, 10.140–144
    off-axis, 10.144–147
    other lens assisted systems, 10.157–158
    other lensless systems, 10.147–148
    reversed wavefront reconstruction, 10.153–157
  microinterferometry, 10.171, 173–190
  noise and speckle patterns, 10.159–161
    techniques, 10.161–163
    unidirectional suppression, 10.163–173
  objective lenses, with/without 10.140–153
  off-axis microscopy, 10.144–147, 157–158
  oil mist, 10.202–206
  on-line microscopy, 10.140–144
  particle analyzing systems, 10.198–201
  particles, 3-D distribution, 10.196–197
  Pawluczyk's design, 10.169–175
  plasmas, 10.196
  principles, 10.104–139
  pulsed and double pulsed lasers, 10.197, 204
  reference beams, collimation, 10.162
  signal-to-noise ratio, 10.161
  speckle patterns, 10.160–162
  stereomicroscope, 10.196
Holography, 6.35; 7.166, 246–256; 8.22; 12.25; 89.2; 94.197–198; 98.362; *See also* Holograms
  content-addressable memory, 89.70–71
  electron holography, 98.331, 362–363, 422–423; 99.171–173, 235–236
    charged dielectric spheres, 99.174, 207–216
    charged microtips, 99.229–235
    double-exposure electron holography, 99.205–207
    electron biprism, 99.173, 176–184
    electron-specimen interaction, 99.174–176
    electrostatic Aharonov–Bohm effect, 99.187–192
    far-out-of-focus holography, 98.373–387
    off-axis electron holography, 98.323–324
    phase-object approximation, 99.174–175, 184–186
    principles, 98.363–373
    reverse-biased *p-n* junctions, 99.172–173, 174, 185, 216–229
    STEM holography, 98.373–387, 422–423
  electron microscopy, 7.167; 96.140–141, 181, 183, 398, 410, 709–712, 753–754, 800–801
  far-out-of-focus holography, 98.373–387, 422–423
  Fresnel holography, 99.173
  history, 89.2–3; 91.259–260, 269, 274, 276–283
  image holography system, 6.127
  imaging techniques, 6.126
  in-line holography, 89.3, 5; 94.198–199, 252
  invention, 96.139, 398
  Japanese contributions, 96.709–712, 753–754
  lensless system, 6.126
  microholographic system, 6.126
  off-axis holography *See* Electron off-axis holography
  optical symbolic substitution, 89.70–71, 74, 77–79
  particle analysis, 10.196–208
    aerosols, 10.198–201, 206

bubble tracks, 10.207
fog, 10.201–202
oilmist, 10.202–206
submicron, 10.206
phase detection, 89.5–6
principles, 89.2–3
scattering amplitude, 94.200–203
sideband holography, 7.167, 168
Holtsmark distribution, 13.220–221
HOLZ *See* High-order Laue zone
Homo-epitaxy, 11.82
Homomorphism:
  defined, 84.312
  Jordan algebra, 84.279–280
  semi-lattice, right linear, 84.69
Homothety, defined, 99.41
Homotopy modification, gradient functions, 99.46, 48–51, 54
Hooge equation, 87.212, 224
Hopfield neural network, 87.2 ff., 87.7, 87.8 ff.
  binary form, 87.18 ff.
  convergence, 87.17
  energy function and, 87.10, 87.16 ff.
  matrix inverse and, 87.23 ff.
  nonbinary form, 87.19 ff., 87.27 ff.
  regularization and, 87.17
  superresolution and, 87.36, 42
H-orbit, 93.9
Hot electron damage, 87.230
Hot hole cascade decays theory, 8.219
Householder transformation, 85.27–28
Howie–Whelan equation, 90.240
HREM *See* High-resolution electron microscopy
Hubble's law, information approach, 90.170, 188
Huffman coding, 82.118; 97.51, 194
Hughes Aircraft Co., 91.199
Human wart virus, 6.260
  reconstruction, 7.370, 371
  three-dimensional reconstruction, 7.323, 368, 370
Humphries eyepiece micrometer, 3.60–63

Hungary:
  electron microscopy, 96.181, 183, 184
  Hungarian Group for Electron Microscopy:
    future developments, 96.191
    history, 96.181, 184
    international congresses, 96.184–186, 189
    Marton, Ladislaus, 96.67–71, 102, 136, 183
Huygens–Fresnel construction, 5.165, 167, 169, 170, 178, 184, 185
Huygens–Fresnel principle, 5.258, 259, 260
Huygens' principle, 86.187
Hybrid entropy, fuzzy sets, 88.252–256
Hybrid multistage filter, 92.40
Hybrid parallel architectures, 87.285–291
Hybrid RT/GT algorithm, 93.25–26
Hydrogen molecule, simulations, 98.68–72
Hydrophilic support film preparation, 8.119–120, 127–130
Hydrophobic grids:
  equilibrium contact angle, 8.97
  water vapor dropwise condensation studies on, 8.80–81, 97–99
Hydrophobic support films, 8.121
  glow discharge treatment, 8.129
Hymenoptera, 5.337; 7.96
Hyperarc, 86.133–134, 137
Hyperbolic cathode-planar anode, electrostatic field, analytical models, 8.223–224
Hyperbolic electron lenses, 1.212
Hypergraph, 86.133–138
Hyperparameters:
  MRF hyperparameters, 97.146–149
  regularization, 97.141–143
Hysteresis error, 83.194–195
Hysteresis thresholding, 88.336

I

IC *See* Integrated circuits
ICCG *See* Incomplete Choleski Conjugate Gradient methods

ICEM *See* International Congresses for Electron Microscopy
Ice nucleation, wet replication study of, 8.99
ICM *See* Iterated conditional modes
Iconic maps, 92.53–55
Iconic memory, 94.289
ICP *See* Inductively coupled plasma spectrometer
IC–SE (software), 13.76, 82
ICSU *See* International Council of Scientific Unions
ICT *See* Integer cosine transform
Ideal:
  abstract algebra, 84.278, 312
  group algebra, 94.25
Ideal data interval, 90.177
Idempotence, 84.275; 99.9
Identity theory, mind-body problem, 94.261
I-divergence, 91.40
IFSEM *See* International Federation of Societies for Electron Microscopy
Ill-conditioned problem, 87.13; *See also* Regularization
Illumination:
  coherence and, 7.141–147
  contrast transfer, 4.75–79
  dark-field, 6.17, 18, 21; 7.148, 273
  in micrography, 7.82
  in mirror EM, 4.207–210
  optical transfer theory, 4.7–9
  photomicrography, 4.389
IM *See* Intensity modulation
Image, 84.208; *See also* Imaging
  additive conjugate, 84.74
  ambiguity *See* Image ambiguity
  analysis *See* Image analysis
  automation of, 3.56
  base, 86.107–110
  binary *See* Binary images
  binary operations between, 84.73–74
  characteristic function, 84.75
  characterization, 2.119, 121, 129, 133
  coding *See* Image coding
  color, 88.298
  on a complex, 84.208–212
  complex, 84.235
  complexity measure, 84.120
  compression *See* Image compression
  constant, 84.74
  continuous, 85.87
  contrast *See* Image contrast
  defined, 84.73; 88.297
  digital *See* Digital images
  enhancement *See* Image enhancement
  faithfulness, 6.58, 59, 88, 92
  feature, 86.116–120, 123–124, 133–134, 137
  formation *See* Image formation
  fractal, 88.237
  fuzzy geometry, 77.256–260
  graph, 84.229
  gray, 88.298
  intensity *See* Image intensity
  mathematical morphology, 84.88
  multiresolution, 86.125, 132
  n-dimensional, 84.210
  normalization, 87.211
  operations between template and, 84.79
  parametric, 84.344
  processing of *See* Image processing
  quality, 86.88, 105, 113, 128; 97.61
  reconstruction *See* Image reconstruction
  recording, 6.30–36
  recovery *See* Image recovery
  reference, 86.105, 107, 129
  remote sensing *See* Remote sensing image
  representation *See* Image representation
  restoration *See* Image restoration
  segmentation *See* Image segmentation
  size, 88.297
  space, 85.86, 117
  speckle intensity, 7.162
  structures, computer simulation, 94.135–140
  thresholding, 84.75
  understanding, 86.90, 114, 142–144, 146, 153, 156, 164, 166

system, 86.90–93, 112, 147, 154, 164
Image algebra, 84.72
  correspondence with:
    mathematical morphology, 84.88
    minimax algebra, 84.85
  first to use term, 84.65
  image processing, 84.64
  minimax algebra properties mapped to, 84.90
  origin, 84.65
  parallel image processing, 90.355–426
    global operations, 90.360–361, 369–371, 424
    image-template operations, 90.361–363, 371–382, 417–420, 424
    pixelwise operations, 90.360, 369, 424
    templates, 90.359–360
Image ambiguity, uncertainty measures, 88.251–260
  grayness, 88.252–256
  spatial, 88.256–260
Image analysis, 2.141; 86.84, 87–88, 92–93; 88.247–296; 97.61–63
  coarse-to-fine, 86.125, 132
  computer analysis, 2.77, 94.2.95, 119; 4.361–383
  contour detection, 88.267
  cooperative, 86.139
  detail enhancement, 92.38–52
    adaptive extremum sharpening filter, 92.44–48
    mathematical morphology, 92.48–52
    multistage one-dimensional filter, 92.39–44
  edge detection, 97.7, 63–64
  equipment, 4.381; 8.36
  expert systems for, 86.87–114
  finite topology and, 84.197–258
  FMAT, 88.269–272, 290
  fuzzy disks, 88.270
  fuzzy segmentation, 88.264–267
  fuzzy set theory, 88.247–291
  fuzzy skeleton, 88.269–272

  inclusions in steel, 4.364–368
  motion analysis, 97.72–74
  pixel classification, 88.267
  plan-guided, 86.125
  problems in, 86.87–89
  process, 86.94, 102, 109
  quantitative, 5.115; 93.222–224; 99.274–285
  stereovision, 97.74–75, 76–78
  strategy, 86.124–126, 129, 132–134, 136, 165
  texture analysis, 97.64–72
  threshold selection, 88.264–267
Image coding:
  algorithms:
    EZW coding, 97.221, 232
    SA-W-LVQ, 97.221–226
  arithmetic coding, 97.194
  digital coding, 97.192–194
  entropy coding, 97.193–194
  Gabor expansion, 97.50–54
  Huffman coding, 97.51, 194
  low-bit-rate video coding, 97.232–252
  optical symbolic substitution, 89.52–59, 68–70
  partition priority coding, 97.201
  predictive coding, 97.51, 193
  regularization, 97.147–148
  standards, 88.2
  still images, 97.226–232
  transform coding, 97.51, 193
  wavelets, 97.198–205
Image compression, 97.192
  applications, 97.50–51, 54
  fingerprint database, 97.54
  JPEG, 97.51, 54
  methods, 97.51
  standards, 97.51, 194
  wavelet transforms, 97.52–53, 194, 198–205
Image contrast, 6.51, 88; 12.48; *See also* Contrast
  in dark-field technique, 5.102, 299, 300–303
  in defocused mode, 5.241

Image contrast, *continued*
  enhancement, 5.102
  in environmental chamber technique, 5.298, 299–303
  in Fraunhofer diffraction, 5.279
  in Fresnel diffraction, 5.272–279, 280
  geometrical approximation, 5.272, 275, 277, 278, 279, 280, 281, 282
  in image transfer theory, 5.287
  inequality criterion, 5.277, 278, 280
  in interference microscopy, 1.84, 85
  international meetings, 96.397–398
  interpretation, 5.241
  for isolated objects, 5.272–279
  for periodic objects, 5.279–282, 284, 285
  in phase contrast technique, 5.102, 299
  in polarizing microscopy, 1.83
  pseudo-classical approximation, 5.265, 266, 272
  semi-classical approximation, 5.246, 249, 265, 266, 272
  stationary phase approximation, 5.265, 266, 273, 279
  for strong objects, 5.273
Image deblurring, 97.155–159
Image discontinuities *See* Discontinuities
Image dissector tracking system, 6.45
Image enhancement, 4.85–125; 88.269; 92.1–3
  binary image enhancement, 92.64–75
    contour chain processing, 92.70–72
    distance transform, 92.72–75
    polynomial filtering, 92.68–70
    rank-selection filters, 92.65–68
  computer system, 4.87–102
  denoising, 97.56–58
  detail enhancement, 92.38–52
    adaptive extremum sharpening filter, 92.44–48
    mathematical morphology, 92.48–52
    multistage one-dimensional filter, 92.39–44
  Gabor expansion, 97.54–55
  gray-scale transformations, 92.4–9
    histogram equalization, 92.6–9
  image fusion, 97.58–59
  line pattern enhancement, 92.52–64
    iconic maps, 92.53–55
    linear filters, 92.55–62
    top-hat transformation, 92.63–64
    topographical approach, 92.62–63
  local contrast enhancement, 92.26–38
    adaptive contrast enhancement, 92.34–36
    background extraction, 92.22–24
    extremum sharpening, 92.32–34
    inverse contrast ratio mapping, 92.30–31
    local range stretching, 92.27–30
    pyramidal image model, 92.36–38
    rank-order statistics, 92.25–26
    shading compensation, 92.20–26
    weighted unsharp masking, 92.24–25
  periodic images, 4.102–112
  resolution by computer synthesis, 4.113–119
  sensor noise, 4.87
  uniformity enhancement:
    adaptive quartile filter, 92.14–16
    center-weight median filter, 92.17
    composite enhancement filter, 92.16–17
    iterative noise peak elimination filter, 92.17–18
Image equation, mirror-bank energy analyzers, 89.403
Image feature frame, 86.117
Image formation, 2.171, 204, 207, 211; 4.1–3
  Abbe's theory, 8.3–4, 14–17; 13.268–272
  angular aperture and, 8.4
  by contrast production, 5.164
  conical illumination, 7.147
  electron, partial coherence and, 7.116–140
  electron microscope, 10.229–233
  ideal, proof of, 13.256–257
  mirror electron microscopy, 94.87–95

in partially coherent illumination, 7.103
process, 5.165, 166, 243, 244
scanning transmission electron
 microscopy, 7.102, 143, 198;
 94.221–231
single lens, 8.3
theory, 13.268–285; 93.174–176,
 215–216
tilted illumination and, 7.152
transmission electron microscopy,
 94.203–213
wave process, 5.243, 244, 245
wave theory:
 Abbe theory, 5.243, 244, 245, 286
 aberration, 5.165, 167, 244, 245, 287
 amplitude contrast, 5.168, 169, 183
 approximations, 5.167–169
 defocusing, 5.165, 245, 287
 exact derivation, 5.167
 first-order approximation, 5.169–181
 Huygens–Fresnel construction, 5.165, 167
 image transform, 5.167
 objective aperture, 5.165, 167
 phase contrast, 5.168
 second-order approximation, 5.183–195
 theory, 5.163, 164–167
Image fusion, image enhancement, 97.58–59
Image intensification, 10.253
Image intensifier, 2.1, 2, 3, 8, 393; 12.71
 cascade, 2.7, 11, 25, 26
 compared with film, 2.14, 17
 focusing of, 2.4, 5, 7
 gain, 2.7, 9, 13, 16, 25
 noise, 2.10, 11, 12, 22
 resolution of, 2.5, 7, 13, 14, 22
 T.S.E.M., 2.8, 9, 10, 11, 17, 24, 25, 27
Image intensity:
 in bright field microscopy, 7.187
 classical, 5.264, 265
 comparison of methods, 5.265, 266
 diffraction integral approximation, 5.263, 264

diffraction theory, 5.260, 261
 equations, 5.266–268
 with Fraunhofer diffraction, 5.263
 with Fresnel diffraction, 5.263
 Fresnel–Kirchhoff integral, 5.260, 261
 Huygens–Fresnel principle, 5.258, 259, 260
 Kirchhoff diffraction integral, 5.260
 phase determination from, 7.190
 reduced equation parameters, 5.266, 267, 268
Image intensity distribution, wave optics, 5.270
Image interchange format, 88.72
Image interpretation, 7.186; 12.26
Image logic algebra, 90.356
Image plan, 86.232
Image plane, monochromatic aberration and, 14.290–294
Image-plane off-axis holography, 7.168; 89.4, 6
 applications, 89.36–47
  crystal defects, 89.44–47
  dynamical phase effects, 89.38–43
  thickness measurement, 89.36–38
 phase distribution, displaying, 89.18–19
 problems, 89.25–35
  hologram recording, 89.32–35
  limited coherence, 89.25–31
  noise problems, 89.31–32
 reconstruction:
  digital, 89.12–18
  light optical, 89.10–12, 21–24
Image processing, 4.153–158; 10.9–14; 13.246–246; 84.338
 Abingdon Cross benchmark, 90.383–389, 425
 abstract algorithm, 86.95, 105
 algorithms, 90.355, 368–425
  Abingdon Cross benchmark, 90.386, 388–389, 425
  fast local labeling algorithm, 90.407, 410–412
  global operations, 90.369–371

INDEX

Image processing, *continued*
  image-template operations, 90.371–382, 417–420
  labeling of binary image components, 90.396–415
    Levialdi's parallel-shrink algorithm, 90.390–396, 415, 425
    local labeling algorithm, 90.400–415
    log-space algorithm, 90.405
    naive labeling algorithm, 90.398–400
    stack-based algorithm, 90.405, 417
  binary image component labeling, 90.396–415
  coherent optical system, 4.129
  computing component geometric properties, 90.415–424
    image-template product, 90.417–420
  consultative system for, 86.86, 94–95, 99, 163
  contrast enhancement, 99.285
  data structures in C computer language, 88.63–108
  design, 88.67
    for flexibility, 88.69
    object-oriented, 88.68
    for portability, 88.71
    for speed, 88.67
  digital system, 4.127–159
    computer system, 4.153–158
    data recording, 4.149–153
    electronic cameras, 4.145–149
    quantization, 4.135–145
    resolution, 4.130–135
  discontinuities, 97.89–91, 108–118
    duality theorem, 97.91, 115–118
    explicit lines, 97.110–115, 154–166
    implicit lines, 97.108–110, 166–181
    line continuation constraint, 97.130–141, 142
  duality theorem, 97.91, 115–118
  edge enhancement, 10.10, 13
  electron off-axis holography, 89.7–47
  error handling, 88.78, 90
  expectation–maximization (EM) approach, 97.127–129
  expert systems for, 86.87–114
  generalized expectation–maximization (GEM) algorithm, 97.93, 127–129, 153, 162–166
  graduated nonconvexity (GNC) algorithm, 97.90, 91, 93, 124–127, 153, 168–175
  gray-scale morphology, 89.326, 336, 366–374
  group algebra, 94.46–50
  hardware, 85.169–172
  high-resolution, 4.86–87
  imaging plate, 99.253, 285–286
  international standard, 88.72
  iterated conditional modes, 97.92
  mathematical morphology, 89.325–389
  operator library, 86.87–88, 93
  optical symbolic substitution, 89.82–84
  parallel/pipelined model, 87.293–294
  parallel processing methodologies, 97.259–297
  predictive coding, 97.51, 193
  problems in, 86.87–89
  scanning electron microscopy, 13.295–298
  SIMD mesh-connected computers, 90.353–426
  spatial frequency filtering, 99.285
  theory, 97.2–3
  three-dimensional, 85.178
  video image, 87.286–287
Image reconstruction, 12.40, 41; 93.110–112; 95.156; 97.86–87, 181–184; *See also* Image recovery; Image restoration
  algebraic reconstruction (ART), 97.160–162
  applications, 97.153–154
    explicit lines, 97.110–115, 154–155
    implicit lines, 97.91, 108–110, 166–181
  blind-deconvolution, 93.144–152
  blind restoration problem, 97.141
  by synthesis of modified projection function, 7.320

INDEX

coherent imaging, 93.152, 160–166
computer processing, 6.258, 259
computer simulation, 93.127–130, 133–134, 162–166
Daugman's neural network, 97.31, 52
deblurring, 97.155–159
digital, 89.5–6, 12–18, 21–24, 33–34, 36, 47
direct, 6.255
discontinuities:
  duality theorem, 97.91, 115–118
  explicit treatment, 97.110–115, 154–166
  implicit treatment, 97.108–110, 166–181
  line continuation constraint, 97.130–141, 142
double-exposure electron holography, 99.207–210
edge-preserving algorithms, 97.91–93, 118–129
  extended GNC algorithm, 97.132–136, 171–175
  GEM algorithm, 97.93, 127–129, 153, 162–166
  GNC algorithm, 97.90, 91, 93, 124–127, 153, 168–175
edge-preserving regularization, 97.93–94
  theory, 97.104–118
electron microscopy, 7.186
electron off-axis holography, 89.5–6, 12–18, 21–24, 33–34, 36, 47
explicit lines, 97.110–115, 154–166
filtered backprojection, 97.160–162
with fluorescence, 14.184–189
from electron micrograph, 6.255
from finite number of projections, 7.327–328
from optical diffraction pattern, 6.255, 256
gray-scale reconstruction, 99.48–49
Hartley transform, 93.140–143
holograms, 99.200–205, 207–210, 218–221
double-exposure electron hologram, 99.207–210
in-line optical bench, 99.200–203, 207
Mach–Zender interferometer, 99.203–205, 219
interferometric, 89.13–16
inverse problem, 97.94–98, 99–101
light optical, 89.5–6, 10–12, 33–34
mathematics of, 14.222–229
missing cone, 14.222–225
non-negative-constrained, 14.226–229
optical diffraction analysis, 7.19, 21, 27
purple membrane, signal-to-noise ratio, 7.215
radiation damage and, 7.192
regularization, 14.222–225; 97.87–89
  Bayesian approach, 97.87–88, 98–104
  discontinuities, 97.89–91, 108–118
  inverse problem, 97.94–98, 99–101
  support-constrained, 14.225–226
  three-dimensional images, 7.304, 314–324, 321–324, 329, 330; 97.59–60
tomographic reconstruction, 97.159–166
two-dimensional images, 7.293–303; 93.131–139, 161–162
of virus protein tube, 6.256, 257
Image recovery; *See also* Image reconstruction; Image restoration
deconvolution with bounded uncertainty
  experiment, 95.240–242
  results, 95.243
  set theoretic formulation, 95.243
optimal solutions, 95.158–159
problem solving, 95.159–160
  convex feasibility problem *See* Convex feasibility problem
  data formation model, 95.156–157, 190–192
  elements required, 95.156
  feasible solutions, 95.159–160, 259–260
  optimal solutions, 95.158–159, 260
  point estimates, 95.158–159
  set theoretic estimates, 95.159–160

Image recovery, *continued*
  solution method, 95.157–158
  projection methods:
    block-parallel methods, 95.216–217
    Browder's admissible control, 95.209–211
    extrapolated method of parallel projections, 95.217–232
    Pierra's extrapolated iteration, 95.211–216
Image representation, 88.65; 97.75, 78–79
  data structures in C computer language, 88.65–66
  Gabor schemes, 97.19–23
    continuous signals, 97.23–30
    discrete signals, 97.30–33
    quasicomplete Gabor transform, 97.34–37
  image analysis, 97.61–63
    edge detection, 97.7, 63–64
    motion analysis, 97.72–74
    stereovision, 97.74–75, 76–78
    texture analysis, 97.64–72
  image coding *See* Image coding
  image compression, 97.192
    applications, 97.50–51, 54
    fingerprint database, 97.54
    methods, 97.51
    standards, 97.51, 194
    wavelet transform, 97.52–53, 194, 198–205
  image enhancement and reconstruction, 97.37, 54–56
    denoising, 97.56–58
    Gabor expansion, 97.54–55
    image fusion, 97.58–59
    image quality metrics, 97.10, 61
    three-dimensional reconstruction, 97.59–60
  joint space-frequency representations, 97.3, 8
    block transforms, 97.11
    complex spectrogram, 97.9–10
    multiresolution pyramids, 97.13–16
    vision-oriented models, 97.16–19
    wavelets, 97.11–13
    Wigner distribution function, 97.9
  machine vision, 97.61–78
  oct-tree, 88.66
  orthogonality, 97.6–7, 11, 13, 22
  pryamid structure, 88.66
  quad-tree, 88.65
  symbolic, 88.66
  theory, 97.2–7
  vision modeling:
    Gabor functions, 97.17, 34–37, 41–45
    sampling in human vision, 97.45–50
    visual cortex image representation, 97.37–41
Image restoration, 87.3, 87.11 et seq., 87.42; 95.156; 99.292–293; *See also* Image reconstruction; Image recovery
  image with bounded noise, bounded versus unbounded noise, 95.246, 248, 251–252
  neural network and, 87.15 ff.
  optimization and, 87.3 ff.
  parallel processing and, 87.6 ff., 87.13
  scanning electron microscopy, 13.295–298
  space variant *See* Space-variant image restoration
  space-variant realization, 99.308–312
  subgradient projections, experiment, 95.253–257, 259
  super-resolution and, 87.2, 11
Images *See* Image
Image segmentation:
  bottom-up, 86.86, 93–94, 111–114, 116, 146, 163
  line-based, 86.114
  region-based, 86.114
  top-down, 86.86, 93–94, 111–112, 114, 116, 146, 163–164
Image of set:
  classical relational calculus, 89.267–276
  fuzzy relations, 89.289–290
  fuzzy sets, 89.284–285

Image-shearing:
  binocular image-shearing microscope
    See Binocular image-shearing
    microscope
  devices for, 3.52–59
  history, 9.224–227
  metrologyh, 9.223–224
Image signal, space-variant image
    restoration, 99.294–295
Image smoothing See Smoothing
Image-template operations, 90.361–363,
    424
  algorithms, 90.371–382, 417–420
Image transfer theory, 5.286–288;
    11.13–15
  contrast transfer function, 5.287
  effect of aberration, 5.287, 288
  effect of aperture, 5.287, 288
  in electron microscope performance,
    5.286
  magnetization ripple, 5.289, 290, 291
  for small deflections, 5.288, 289
  for thin polycrystalline films, 5.289, 290,
    291
  wave aberration term, 5.287, 288
Image transforms, optical diffraction
    analysis, 7.19
Imaging See also Image
  applications, medical, 85.79
  atomic imaging:
    history at international meetings,
      96.402–403, 408
    Japanese contributions, 96.615–618,
      623, 625–626
    processing by fast Fourier transform,
      96.622–623
  biological specimens, 7.186
  crystal-aperture scanning transmission
    electron microscopy, 93.58–59,
    63–66, 87–90
    gold adatoms, 93.94–100
    subatomic detail, 93.100–105
    double-passage coherent imaging,
      93.152–154
  efficiency, partially coherent

    illumination and, 7.135
  error handling, 88.78, 90
  with fluorescence:
    delayed, 14.191–196
    lifetime, 14.199–204
    multiwave three-dimensional, and
      image reconstruction, 14.184–189
    quantitative, 14.183–184
    ratio, 14.180–182
  incoherent, 7.271
  of linear lattices, 1.116
  magnetic microstructure, 98.328–333
  multispectral processing, 93.313–319
  on-line, 11.49, 50
  optics See Imaging optics
  orientation analysis, 93.220–232,
    323–326
    algorithms, 93.319–320
    applications, 93.300–319
    automation, 93.320–322
    domain segmentation, 93.278–299
    edge detection operators,
      93.231–239
    image acquisition, 93.228–231
    image analysis, 93.219–228
    image processing, 93.231–239
    image resolution, 93.275–276
    intensity gradient operators,
      93.246–287
    presentation of results, 93.239–244
    quantitative parameters, 93.244–246
  phase retrieval:
    blind-deconvolution problem,
      93.144–145
    coherent imaging through turbulence,
      93.152–166
  resolution:
    crystal-aperture scanning transmission
      electron microscopy, 93.91–94
    orientation analysis, 93.275–276
    subatomic, 93.64
  of single atoms, 1.124
  stereo imaging, 11.154; 91.253–255
  theory, 10.16–34, 224
  Z-contrast imaging, 90.289–293

Imaging optics:
  aperture, 6.13, 14
  darkfield transillumination, 6.17, 18, 19
  depth of field, 6.14
  dipping objective, 6.15, 16, 17
  fluorescence, 6.18, 19
  high-dry objective, 6.14, 16
  K-mirror, 6.13, 14
  mirror objective, 6.19, 20, 21
  object plane positioning, 6.13
  photomicrography, 4.388–389
  reflection objective, 6.17
  resolution, 6.14
  water-immersion lens, 6.15
  water-immersion objective, 6.16, 17
Imaging plate, 99.241–242, 248–250
  CBED pattern with, 99.269–270
  computed radiography and, 99.242, 263–265
  dynamic range, 99.269–274
  electron diffraction with, 99.270–274
  erasing, 99.253
  exposure, 99.250–251
  fading, 99.258–259
  granularity, 99.259–262
  high-resolution electron microscopy with, 99.274–285
  image processing, 99.253, 285–286
  quantitative image analysis, 99.274–285
  radio luminography and, 99.242, 263–265
  reading, 99.251–253
  resolution, 99.257
  RHEED and, 99.286–288
  sensitivity, 99.254–257, 262, 265–269
  transmission electron microscopy and, 99.262–263, 265–269, 288
Imaging system, mirror EM, 4.210–211
Immersion lens, 10.244, 245; 12.94; 13.92–107
Immuno-cytochemistry, 6.212
Immunoelectron microscopy, Japanese contributions, 96.740
Immunofluorescence microscopy, 10.316–317

Immuno-histochemical technique:
  antiserum reagent, 6.218
  coupling reagent, 6.218
  diffusion problems, 6.218
  disadvantages, 6.218
  with enzyme marker, 6.218
  with horseradish peroxidase marker, 6.218, 219
  HRP-specific immunoglobulin reagent, 6.218
  immunoglobulin specific reagent, 6.218
  limitations, 6.219
  specific site coupling by HRP, 6.218, 219
Impact ionization, 99.81–82
Impedance boundary condition, three-dimensional microstrip circulators, 98.288–301
Imperfect data, methods of handling, 94.151, 194
Implantation, 83.29, 110
Implication operator, fuzzy relations, 89.277–279
Implicit lines:
  image processing, 97.91
  image reconstruction, 97.108–110, 166–181
IMPRESS (software), 86.105, 107, 109, 112
Improved phase-object approximation, 93.190–192
Impurity, periodic distribution of, 82.215
Inaccuracy measure, 91.37, 38
Incidence relation, 84.207
Incident beam, energy spread, 7.160
Incident cells, 84.207
Incident subcomplexes, 84.219
Incident wave, 90.208
Incoherent dark field illumination, 7.148
Incoherent tunneling, 89.128
Incomplete Choleski Conjugate Gradient (ICCG) methods, 13.10
Indanthrene, diffraction pattern, 4.119
Indefinable set, 94.159, 166
Index of anisotropy:
  domain segmentation, 93.298, 299

orientation analysis, 93.223, 245–246, 324, 325
Index of refraction *See* Refractive index
Indiscernibility, 94.152–157
Indium, 91.144
Indium aluminum arsenide, 86.50
Indium antimonide, semiconductor history, 91.173–175, 179
Indium arsenide, 86.50
Indium gallium arsenide, 86.50
Indium phosphide, 86.50
    avalanche phosphodiodes based on, 99.142–146, 156
    ionization rates, 99.116–118
Indium phosphide/indium gallium arsenide avalanche photodiodes:
    SACGM, 99.73, 92–94, 96–156
    SAM and SAGM, 99.90–92
Indium stibnite, properties, 92.82
Inductively coupled plasma spectrometer (ICP), 9.292
Inelastic electron scattering, 2.180; 3.180–183, 190, 201, 207–211; 9.68–82
    angular conditions and, 9.71–72, 91–94
    annular detector aperture and, 7.271
    coherence and, 7.122
    contribution in dark field and bright field images, 7.228
    defined, 9.70–74
    field electron emissions, in, 8.220
    image contrast and resolution and, 7.218
    instruments, 9.68–70
    mean free path, 9.72
    object wave function and, 7.211
    partially integrated cross-section, 9.72
    phase determination and, 7.191, 274
    phase object approximation and, 7.195
    quantum electron transport, 89.134–135
    radiation damage and, 7.263
    total cross-section, 9.72
    weak phase approximation and, 7.268
Inertial navigation, 85.2–16, 54, 57, 60
    error description, 85.11–16
    support, 85.2, 19, 24, 53–54, 61, 65–66

Inessential terms, maxpolynomials, 90.102–103
Inference, 86.83
    classical, 89.316–317
    deductive, 86.146
    engine, 86.83
    fuzzy, 89.317–318
    method of cases, 89.322–323
    modus ponens, 89.316, 318–320
    syllogism, 89.322
Inf-generating mapping, 89.378
Infinitely thin element, 86.228
Infinite processes, minimax algebra, 90.75–84
Infinitesimal operator, 84.136, 84.138, 84.146, 84.184 ff.
    dilations, 84.152
    rotations, 84.152
    smooth deformations, 84.154
Infinitesimal transformation, 84.183
Influenza virus, 6.237
Information *See* Fisher information; Physical information
Informational uncertainty, 97.10
Information divergence, 90.144–145, 194–196
Information flow rate, 90.173
Information function, 94.153
Information measures, 91.37–41
    unified r,s, 91.41–132
Information processing:
    biological, 94.260, 262–263, 264, 292
    chaotic dynamics, 94.289
Information radius, 91.37–38
    unified r,s, 91.113
    M-dimensional, 91.76–77
Information retrieval, 89.306–307
    fuzzy relations, 89.272, 307–310
Information storage, magnetic materials, 98.324
Information tables, 94.152–154
Information technology, 91.150, 190
    impact on the future, 91.210–212
    unlimited database, 91.210

Information theory:
  EPR experiment, 94.302–304
  statistical, 91.37–41, 110–132
Infrared camera, 12.347
Infrared catastrophe, 90.179
Infrared microscopy, 10.8, 15
Infrared radiation, 83.29
Infrared scanning technique, 12.351
Inhomogeneities, use of mirror EM
  electrical surface, 4.242–246
  geometric, 4.241–242
  magnetic, 4.247
Inkjet printer, 91.237
In-line holography, 89.3, 5; 94.198–199, 252
Inner potential, 12.19
Input, single-electron devices, isolation, 89.238–240
Input reflection coefficient, 92.181
Insects:
  flight muscle, 6.193
  radiation damage, 5.337
In situ specimen treatment, 11.75
In situ sublimation, 11.76
Institute of Physics, Electronics Group, 91.139–140, 260, 269, 270
Instrumentable quantizers, 84.4
Insulating materials, 83.22, 24
  beam-induced conductivity, 83.234
  coating, metal, 83.219, 233
  conductive chemical treatment, 83.233
  mirror electron microscopy, 94.96–97
  photoelectron microscopy, 10.300–305
  scanning electron microscopy, 83.208, 219
  scintillators, 83.225
Integer cosine transform (ICT), 88.10, 25
  derivation:
    order-8, 88.31
    order-16, 88.42–44
  fast computation algorithm, 88.57–58
  fixed-point error performance, 88.52–56
  implementation, 88.49–52
  performance, 88.44–49
Integral equation, 86.193
  Fredholm equation, 86.258, 260
  lenses, 13.153–154
  magnetic fields, 13.169–173
  in variations, 86.246, 258
Integral transform, 84.134
  condition for invariance, 84.135
  condition for uniqueness, 84.135
  covariant, 84.161, 164
  invariant in the strong sense, 84.146–151, 155
  kernel, 84.154
  with respect to dilations and rotations, 84.153
Integrated circuits, 12.140; 83.14, 67, 76; 87.233, 245; 91.146–169
  history, 91.181
  images, 10.4, 6, 7, 12, 13
  planar, 91.147–148, 195
  quantum-coupled architectures, 89.217–243
  quantum devices, 89.183–184, 208
  scanning electron microscope, 13.126, 153
  T-structure transistors, 89.183–184
  ultralarge-scale integrated chips (ULSI), 89.208–210, 215–217
  United Kingdom, 91.150–151, 168–169
  very high-speed integrated circuits (VHSIC), 89.208–210, 215–217
Integrated digital network, 91.197
Integrated transforms, signal processing, 94.320
Integrating microdensitometry, 6.135
Integration transformation:
  electrostatic lenses, 97.396–403
  Glaser's bell-shaped magnetic field, 97.393–395
  magnetic lenses, 97.369–381, 389–392
Intel Corp., semiconductor history, 91.169, 195
Intelligence; *See also* Animal intelligence
  gradations of, 94.269–270
Intensity, light definition, 10.105, 107
Intensity coding, optical symbolic substitution, 89.58, 77
Intensity gradient, 93.231

INDEX

Intensity gradient analysis:
  algorithm, 93.319–320
  boundaries, 93.267–272, 324
  flow charts, 93.228, 230, 300, 302
  image resolution, 93.275–276
  intensity gradient operators, 93.246–272
  with multispectral analysis, 93.301, 302
  noisy images, 93.257, 277
  pixels:
    numbering, 93.233, 268, 284
    rectangular aspect ratio, 93.272, 274–275
  with porosity analysis, 93.300, 301
  statistical analysis of data, 93.278–284
  three-dimensional, 93.284–287
Intensity modulation (IM), 99.68
Intensity transmittance, 89.33, 34
Intentionality, 94.294
INTERAC (software), 13.217
Interaction loop, 85.87
Interconnect capacitance, 89.209
Interconnectless architecture, 89.219, 224
Interdigital electrode, mirror electron microscopy image, 4.255, 266
Interdomain boundaries, ferromagnets, 87.176–180
Interface condition:
  of eddy current field, 82.21, 22, 27, 29, 30, 35, 55, 57, 59
  for magnetic scalar potential, 82.15, 50, 51
  treatment as boundary condition, 82.32, 36
Interface states, electrical noise, 87.228, 230, 234, 245
Interfacial tension, 8.111
Interference:
  echographic image processing, 84.326
  electron biprism, 12.13, 15, 19
  electron wave devices, 89.103–110
Interference contrast, 1.68, 83
Interference field, 12.57
Interference fringes, 8.34
  Aharanov–Bohm effect, 5.253
  domain wall images, 5.254, 255, 293
Interference microscope:

Dyson, 1.68, 80
extinction factor of, 1.82
Leitz, 1.68, 80
polarizing, 1.80
spatial resolution, 1.87, 111
systematic errors of, 1.100
uses of, 1.99
Interference microscopy, 5.102; *See also* Holographic microscopy
  advantages, 6.104
  applications, 3.67, 81; 6.105; 10.81
  basis of, 10.112
  DIC, 10.178
  diffraction effects, 10.79
  double exposure, 10.183–185
  heterodyning, 10.82, 83
  multiple beam, 10.79–80
  noise reduction, 10.186
  phase objects, 10.186–190
  plane and spherical waves, 10.114
  real-time, empty wavefront, 10.179–181
  resonant microscope, 10.79–80
  scanning systems, 10.76–81
  shearing, 10.181–183
  of solids, 10.195
  thin film circuits, 10.81
  two-plane waves, 10.109–114
  two-spherical waves, 10.115–117
  uniform interference, 10.113
Interferometry, 10.58, 113, 176
  amplitude division interferometry, 99.193–194
  deflection-sensing, 87.192–193
  digital reconstruction, 89.13–16
  double quantum wire Aharonov–Bohm interferometer, 89.172–178, 180, 186, 199
  holography and, 10.158
  length measurements, 8.49
  Mach–Zender interferometer, 1.68, 80; 10.113, 176; 12.40; 89.11, 146–149; 99.203–205, 219
  Michelson interferometer, 6.111–114; 8.29; 10.113, 176; 89.77
  optical symbolic substitution, 89.77
  radioastronomy, 91.286, 287, 289

Interferometry, *continued*
  retardation, 6.136
  scanning interferometer, 8.26, 33
  speckle interferometry, 93.143–144
  triangular interferometer, 10.158
  two-beam interferometers, 8.34
  wavefront division interferometry, 99.194–200
  wavelength measurement, 8.39
Intermediate node, 90.39
Intermolecular pair potentials, additivity, 87.54–55
Internal friction, 87.224
Internally undefinable partition, 94.169–170
International Assembly for Electron Microscopy, 96.6
International Committee for Electron Microscopy, 96.6, 12–16
International Congresses for Electron Microscopy (ICEM), 96.33, 385
  biological research history, 96.387–391
  history, 96.386–387
    electron optics research, 96.405–412
    materials science research history, 96.395–403, 408
  instrument exhibitions, 96.34, 394–395
  oversight, 96.32–33
  proceedings, 96.403, 406, 408, 411
International Council of Scientific Unions (ICSU), association with IFSEM, 96.5–7, 9–10
International Federation of Societies for Electron Microscopy (IFSEM):
  constitution, 96.8–10, 18–19, 21, 32
  Executive Committee, 96.22–23
  General Assembly, 96.22, 33
  General Secretary, 96.23, 25
  growth, 96.9, 25, 27, 34–35
  industry relations, 96.33–34
  international congresses, 96.32–33
  International Council of Scientific Unions, association, 96.3, 5–7, 9–10
  Joint Commission, rise and fall, 96.5–7
  membership, 96.30
  member societies, 96.25, 27, 30, 34
  objectives, 96.21
  origin, 96.3–5, 155–156
  presidents and secretaries, 96.9, 20, 22, 24
  proposal by International Committee, 96.7, 16–18
  regional committees, 96.28–29, 31–32
  responsibilities, 96.10–11
International Symposium on Nanostructure Physics and Fabrication, 91.215
International Telephone and Telegraph Consultative Committee, 91.201, 204
Interphako system:
  applications, 6.108, 109
  color phase contrast, 6.107
  construction, 6.105, 106
  field interference, 6.110
  fringe interference, 6.106, 110
  interference images, 6.106
  optical system, 6.106
  phase contrast, 6.107
  phase interference contrast, 6.107, 108, 109, 115
  shearing interference, 6.107, 110
Intersection:
  τ-mapping, 89.339–343, 376–377
  three-dimensional display, 85.126, 130, 158
Intracellular organelles, 6.183
Intrascope, 5.16
Intravital microscope, 7.73
  adaptations, 6.2
  animal table, 6.7–13, 46
  applications, 6.2, 46
  dark-field image, 6.17, 18, 19
  design, 6.2–7, 46
  fluorescent image, 6.18, 19
  illumination, 6.2, 21–30
    epi-illumination, 6.4, 9, 11, 18, 28–30
    fiber optic illumination, 6.2, 10
    transillumination, 6.4, 9, 11, 13, 17, 18, 21–28

image recording system, 6.4, 30–36
imaging optics, 6.13–21
imaging system, 6.2, 46
light pipe illumination, 6.2, 4, 9, 10, 26–28
physiology study, 6.1, 2, 45, 46
problems, 6.38
servo-controlled focusing, 6.2
specimen:
   life support system, 6.2, 7, 9
   manipulation, 6.2, 7
   stage, 6.2, 5
   telescopic optical system, 6.6
   transfer lens, 6.6, 7
   vibration, 6.4, 5
Intrinsic aberrations, to ninth-order approximations, 91.28–30, 34
Invariance, 84.131, 133, 134, 147, 149
Invariant coding, 84.132, 136, 167
Invariant functions, 84.155, 157
Invariant pattern representation, Lie groups theory, 84.131–192
Invariant recognition, 84.131, 142
Invariants, brain function, 94.292–293
Inverse contrast ratio mapping, 92.30–31
Inverse filter, 82.122
Inverse mobility tensor, 92.87
Inverses:
   group algebra elements, 94.22–27
   matrices with group algebra elements, 94.42–43
Inverse Toeplitz covariances, 84.271–272
Inversion:
   crystal structure factors, 90.257–265
   matrix, 85.21–22, 37, 43, 50
Inversion formulas, 94.332–333
Inverting inequality, 90.30–31
Ion beam milling, 83.45
Ion-emission microscopy, 11.121–129
Ion etching, Japanese contributions, 96.784
Ionic forces, 87.102–112
   characteristic separation length, 87.104
   contact value theorem, 87.103, 108
   Derjaguin–Landau–Verwey–Overbeek theory, 87.110

diffuse counterion atmosphere, 87.102, 107–108
"double layer," 87.102
electrolyte solution effect, 87.106–112
Poisson–Boltzmann equation, 87.111
probe-sample charging in ambient liquids, 87.102–106
repulsive ionic pressure, 87.108–109
two-slab ionic pressure, 87.103–104
unwanted, 87.110–111
"weak overlap approximation," 87.108
Ionic transport mechanism, histochemical study, 6.194
Ionization:
   photodiode absorption layer, 99.110–113
   in semiconductors, charge and grading layers, 99.113–115
Ionization energy, 11.107
Ionization rates, indium phosphide, 99.116–118
Ionizing radiation, 87.231, 245; *See also* Radiation damage
Ion microprobe, 6.276; 10.220; 11.128, 129–140
Ion potential, 86.181, 221
Ion probe mass spectrometry, 6.276
Ion probe microscopy, 11.101–103
   ion sources, 11.103–112
   mass separation of primary ions, 11.113–120
Ions, vacuum microelectronics, 83.7, 8283.63
Ion sputter gun, 11.61, 76
Iron, fine magnetic particles, 98.408–409
Iron–carbon phase diagram, 10.295
Iron-shrouded solenoid electron lens, 5.218–220
Irons pole piece electron lens, 5.224, 225
Irradiance, defined, 10.105, 107
Irregular-geometry circulator, 92.183
Irreproducibility, quantum-coupled devices, 89.222, 244
Isocon tube, 9.40, 42–43, 44, 60
Isogyre, 1.51, 53
Isolated node, 90.50

Isolated zero, 97.224
Isometry, 84.281
Isomorphism:
　embedding minimax algebra into image algebra, 84.85
　finite abelian groups, 93.6–9
　labelled subgraphs, 84.231
　subgraphs, 84.229
Iso-phase axis, 94.232, 235, 236, 240, 241
Isoplanatic imaging, 12.28
Isotopy subgroup, 93.9
Isotropic intrinsic aberrations, 97.360–361
Isotropic operator, 93.235, 236, 266, 267, 269, 277
Italy:
　electron microscopy:
　　biomedical sciences, 96.202–204
　　fixation and staining, 96.203
　　research, 96.202
　　specimen stages, 96.205–206, 208, 213
　　superconductors, 96.213
　electron microscopy laboratories:
　　Ispra, 96.199
　　Rome, 96.193–194, 202, 212
　　University of Bologna laboratory, 96.194–206, 209–211
　Italian Society of Electron Microscopy (SIME):
　　founding, 96.198–199
　　international congresses, 96.208–209, 213
　　publications, 96.861–876
　　research, 96.202
Iterated conditional modes (ICM), image processing, 97.92
Iteration scheme, phase problem solution and, 7.226
Iterative algorithms, 93.110–111, 167
　blind-deconvolution problem, 93.145, 148
　interative algorithm B, 84.287, 302
Iterative noise peak elimination filter, 92.17–18
Iterative transformations, neural networks, 94.278–284
Ito, Kazuo, 96.659

J

Jacobian, 84.291
Jacobian determinant of Schrödinger equation, 13.251–254
Japan:
　electron microscopy:
　　accelerating voltage elevation, 96.707–709
　　anticontamination devices, 96.697–698
　　atomic imaging, 96.615–618, 623, 625–626
　　bacterial ultrastructure, 96.735, 741–744
　　biological specimen preparation, 96.725, 727–729, 733
　　convergent-beam electron diffraction, 96.766
　　critical-point drying method, 96.783
　　cryoelectron microscopy, 96.739
　　crystal growth experiments, 96.603–604, 608–610, 698–700, 751–752
　　cytology, 96.682
　　diffraction contrast experiments, 96.601–602
　　electron diffraction, 96.599–600, 693–694
　　electron lens aberrations, 96.685–688, 705–707
　　electron microscope development, 96.599–601, 679–680
　　electropolishing methods, 96.750–751
　　field-emission gun, 96.710–713
　　gas evaporation method, 96.785–786
　　high-resolution electron microscopy, 96.759–765
　　high-voltage electron microscopy, 10.254; 96.612–615, 700–704, 736–737, 754, 756–759

holography, 96.709–712, 753–754
immunoelectron microscopy, 96.740
ion etching, 96.784
materials science, 96.619–620, 622, 749–768
microbiology, 96.681–682
microcharacterization, 96.766–768
microgrid development, 96.775–778
nanoprobe analytical electron microscope, 96.713–716
rapid freezing techniques, 96.740–741, 783–784
replicas, 96.774–775
scanning electron microscopy, 96.704–705, 712–713
scanning tunneling microscopy, 96.719, 721
specimen manipulation, 96.694–696, 773–774
spin-polarized SEM, 96.716–718
thin films, 96.611, 614
thinning methods, 96.750–751, 781–783
ultrahigh-vacuum microscope, 96.716
ultramicrotomy, 96.778–781
ultrathin sectioning, 96.725
virus ultrastructure, 96.682–683, 735–741
wet-cell microscopy, 96.786–787
X-ray microanalysis, 96.696
yeast ultrastructure, 96.744–745
Electrotechnical Laboratory:
electron lens contributions, 96.258
instrumentation, 96.257–258
specimen temperature research, 96.258, 260–261
history of early research:
cradle period, 96.232–234
improvement and reformation period, 96.235–236, 239–241
period of application and newcomers in manufacturing, 96.241–243
microscopes, commercial production, 96.217, 226, 243, 257–258, 662–664
semiconductor industry, 91.146, 149, 168–169
Japanese Society of Electron Microscopy (JESM):
founding, 96.217, 723
historical activities in electron microscopy, 96.223, 724
honorary members, 96.225
membership, 96.225
papers submitted to meetings, growth, 96.724–725
presidents, 96.224, 227
publications, 96.225–226, 627, 852–877
Ronbun Prize, 96.222, 226
Seto Prize, 96.218–221, 226, 732
symposia themes, history, 96.730–731
Japanese Society for the Promotion of Science, 37th Subcommittee:
aberration calculations, 96.232, 235, 243
establishment:
background, 96.217, 227–228
committee members, 96.229
first meeting, 96.229–231, 723
prospectus, 96.229–230
research subjects, 96.230
industrial research, 96.240–241, 243
microscope construction, 96.233–236, 238–239
policies, 96.231–232
stabilization of high tension, 96.233, 235
training of young researchers, 96.240
translation of von Ardenne's book, 96.240
World War II impact, 96.234, 241–242
J-divergence, 91.37, 40, 71–79, 113, 131–132
Jeffreys invariant, 91.37
Jellium model, 82.199, 207, 208, 219
Jena IMA 10 laser microanalyzer, 9.253–263, 266
autocollimation beam path, 9.258
auxiliary spark gap, 9.258–260
with EM 10 measuring apparatus, 9.267
observation in polarized light, 9.257

Jena IMA 10 laser
 microanalyzer, *continued*
 with optical multichannel analysis,
 9.282–283
 optical path:
 incident illumination, 9.257
 transmitted light, 9.253–255, 257
 photomicrography, 9.257
 projection of electrodes by imaging
 condenser, 9.258
 scanning stages, 9.266
 vacuum specimen chambers,
 9.260–263
Jensen difference divergence, 91.38
Jensen inequalities, 91.46, 63, 64, 82–87,
 100, 106, 110, 131
JEOL Ltd.:
 DA-1, 96.660
 founding, 96.659–660
 JE-100B, 96.714
 JEM-5 reflection electron microscope,
 96.694
 JEM-T1, 96.661–662
 200CX, 10,257
JESM *See* Japanese Society of Electron
 Microscopy
JFET *See* Junction field-effect transistor
Jodrell Bank Observatory, 91.286–287,
 289
Johnson limit, 89.243
Johnson noise, 87.206
Joint optimization, 82.167
Joint space–frequency representations,
 97.3, 8
 block transforms, 97.11
 complex spectrogram, 97.9–10
 multiresolution pyramids, 97.13–16
 vision-oriented models, 97.16–19
 wavelets, 97.11–13
 Wigner distribution function, 97.9
Jordan algebra, 84.262, 263
 defined, 84.311
 dimension, 84.275–276
 generation, 84.274
 homomorphism, 84.279–280, 312
 multiplication tables, 84.312–313

simple, 84.276–277, 280
special, 84.312
symmetric linear mapping, 84.273
symmetric product, 84.272–273, 311
Jordan theorem, 84.198, 212
Joseph algorithm, 85.21
Joule heating, 83.63
Jover and Kailath algorithm, 85.39
JPEG image compression, 97.51, 54
Jugoslav Iskra, microscope production,
 96.484–485
"Jump-to-contact" phenomenon, 87.128
Junction field-effect transistor (JFET),
 86.32; 87.225, 227, 231, 245
Junction finding, 88.337
Junction transistor, 91.143, 144

K

Kadanoff–Baym–Keldysh formalism,
 89.118–119, 134–135
KA equipment, 6.82–86
Kalman filter, 85.2
 adaptive algorithm, 85.24–25
 backward algorithm, 85.49–54, 57, 69
 forward algorithm, 85.49–50, 54, 69
 gain matrix, 85.20–21, 39, 42
 Kalman–Bucy algorithm, 85.17–22
 space-variant image restoration, 99.292,
  293, 295
  estimation algorithm, 99.297–298
  state-space representation,
   99.295–297
  steady-state solution, 99.299
 square-root algorithm, 85.22–24
Kalman filtering, navigation, 85.53–70
Kane band tails, 82.232–234
Kanning acoustic microscope, 11.154
Karhunen–Loéve expansion, 94.319
Karhunen–Loéve transform (KLT), 88.7
Karnaugh maps, 5.80, 83, 84; 89.66
Karp algorithm, 90.61–63
Kazato, Kenji, 96.659, 660, 661
KD*P crystal, 10.87, 88
k-d tree, 86.147
Kelvin equation, 87.120

Keratin, 5.300
Kernel constraint, 89.333
Kernel representation, 89.330–331, 359
KFA equipment:
  applications, 6.72, 73, 74, 75, 76, 77
  for biological microscopy, 6.71
  comparison with KFS device, 6.77, 78, 79
  construction, 6.71
  with high-resolution cinemicrography, 6.77
  image contrast, 6.72
  for immersion refractometry, 6.77
  for living cell study, 6.74, 77
  sensitivity, 6.71, 72
  for tissue culture study, 6.77
KFS equipment, 6.77–81
Khalimsky space, 84.210
Kikuchi, Seichi, 12.3–5
Kinetic energy, 82.209
Kinetic energy of localization, 82.234, 239
Kirchhoff diffraction integral, 5.260, 272
Kirchhoff–Fresnel integral, 93.197, 207
Kirchhoff's current law, 89.115
Klauder's theory, multiple scattering, 82.200, 223–230
Kleene–Dienes operator, 89.278
Klein–Gordon equation, 90.154, 160–161, 164, 179, 185–187; 92.88
  charged-particle wave optics, 97.259, 276, 337, 338
  Feschbach–Villars form, 97.263, 322, 339–341
Klein paradox, 95.328–330, 332
K-mirror, 1.15, 32, 34, 38
Knoll, Max, 83.204; 95.13–14, 16–18, 39, 42; 96.132–135, 142
Knowledge:
  acquisition, rough set theory, 94.184, 191
  declarative analysis, 86.81–86, 89, 91, 93, 144, 146, 154, 163–164, 166, 188
  image processing and analysis, 86.91–92, 144–145
  procedural knowledge, 86.82, 85, 87, 89, 146
  representation, 86.87, 89, 149
    algebraic, 86.86, 149–150, 153, 166, 167
    in computer vision, 86.84–86
    in logic, 86.143, 166–167
    network, 86.117–118, 135, 157
    scene domain knowledge, 86.91–92, 144–145
Knowledge base, defined, 86.83
Köhler illumination, 6.25, 26, 53; 7.141; 10.7; 11.131
Kolmogorov's addition axiom, 94.171
Kossel, Walter, 12.2, 3–4
Kossel lines, 12.2
Kösters prism, 8.44
Kramers–Kronig relation, 87.56
Krypton, melting point, 98.10
Kuipers test, domain segmentation, 93.283
Kullback–Leibler cross-entropy, 90.144, 145; 91.60
Kymography, projection, 7.94
Kyoto University:
  atomic imaging, 96.615–618, 623
  crystal growth experiments, 96.603–604, 608–610
  cytology, 96.682
  diffraction contrast experiments, 96.601–602
  electron diffraction camera development, 96.599–600
  electron microscope development, 96.599–601, 679–680
  high-voltage electron microscopy, 96.612–615
  microbiology, 96.681–682
  thin films, 96.611, 614
  virology, 96.682–683

L

Labeling:
  binary image components, 90.396–415

Labeling, *continued*
   fast local labeling algorithm, 90.407, 410–412
   local labeling algorithm, 90.400–415
   log-space algorithm, 90.405
   naive labeling algorithm, 90.398–400
   stack-based algorithm, 90.405, 417
finite topology and image analysis, 84.210, 224, 230
LACBED *See* Large-angle convergent-beam electron diffraction
Lagrange–Helmholtz relation, 8.145
Lagrange mechanics, 91.4
Lagrangian coordinates, 86.260
Lambert's law, 8.171
Landauer formula, 89.117, 119
Landau–Lifshitz principle, 98.325
Langmuir relation, 11.107; 87.104
Lanthanum hexaboride electron gun, 9.191; 10.248, 250; 11.4; 93.93
Laplace fields, rotationally symmetric, 8.228
Laplace filter, 10.10
Laplace pressure, 87.121
Laplace's equation:
   curved electrodes, 13.9–10
   discretized form in FDM methods, 13.7–8
   electrostatic field disturbance and, 86.250, 252, 256, 263; 87.130–131
   and field emission sources, 13.152
   magnetic field and, 12.124; 13.25; 87.131, 140
   in model fields, 13.144
   in numerical methods, 13.154, 156
   in potential theory, 13.135–136
Laplace transform, 85.16, 60
Laplacian pyramid, image compression, 97.51–52
Lapped orthogonal transform, 97.11
Large-angle convergent-beam electron diffraction (LACBED), 90.238, 240–241
Laser *See* Lasers

Laser atomic absorption, 9.288–292
Laser CT microscope, optics, 14.230–232
Laser-diode feedback system, 87.193
Laser microanalysis, 9.243–318; *See also* Jena LMA 10 laser microanalyzer
   accuracy, 9.287, 318
   applications, 9.287, 293–299, 317, 318
     archaeological, 9.285, 299, 318
     forensic science, 9.284–285, 299, 318
     medical, 9.285, 299
     metallography, 9.284
     metallurgy, 9.284, 293, 299, 318
     mineralogy, 9.284, 293, 318
     silicate technology, 9.284, 299, 318
   area analysis, 9.246, 287, 318
   autocollimation beam path, 9.258
   auxiliary spark gap, 9.258–260, 286
   commercially manufactured microanalyzers, 9.272–275
   control apparatus development, 9.286
   detection sensitivity, 9.286, 287, 316, 318
   distribution analysis, 9.246, 287, 318
   economic aspects, 9.317, 318
   efficiency, 9.299, 316–317
   electronic area detectors, 9.280–283
   element specificity, 9.286, 299, 316
   energy measuring apparatus, 9.266–270
   laser:
     glass, 9.247
     ruby, 9.247, 250–253
     YAG laser, 9.247, 250
   lateral spatial resolving power, 9.317
   layer analysis, 9.246
   line analysis, 9.246, 287, 318
   local analysis, 9.245–246, 287
   microatomic absorption, 9.288–292
   microchemical method, 9.244–245
   microemission spectral analysis, 9.245, 292
   micromass spectrography, 9.287–288
   microplasma spectrometry, 9.292
   microprobe mass analyzer, 9.288
   microscope equipment, 9.253–266
   nonconductor analysis, 9.284, 285

observation:
   polarized light, 9.257
   under microscope, 9.285
optical media, 9.247
optical path of transmitted light, 9.253–257
"optical pumping," 9.247
Q-switching:
   electro-optical switches, 9.248–249
   rotating reflectors, 9.249
   saturable absorbers, 9.249–253
   solid-state lasers, 9.286
qualitative analytical technique, 9.278
quantitative analytical technique, 9.278–279, 318
radiation detectors, 9.271–284
recording:
   photoelectric, 9.279–283
   photographic, 9.257, 278–279
reproducibility, 9.286, 287, 316, 318
resonator, 9.247
sample:
   preparation, 9.284, 285, 317
   size, 9.285
   transparent, 9.286
scanning stages, 9.263, 265–266
spectral apparatus, 9.271–284
spectrographs, 9.271, 276–278
spike number measuring apparatus, 9.266–270
technique, 9.247–283
vacuum chambers, 9.260–263
volume analysis, 9.246, 287, 318
Laser microemission spectral analysis, 9.245
   with infrared microabsorption spectrophotometry, 9.292
Laser micromass spectrography, 9.287–288
Laser microplasma spectrometry, 9.292
Laser microprobe mass analyzer, 9.288
Laser probe emission spectrography, 6.276
Lasers, 1.69; 87.246
   beam intensity fluctuations, 10.10
   beam photocurrents, 10.66–67
   chopped laser illumination, 10.58
   continuous wave, data, 10.148
   frequency-stabilization, 8.34
   fusion experiments, 10.196
   gallium arsenide, 10.84
   gas laser, 91.179
   glass laser, 9.247
   gyro strapdown system, 85.65
   heterodyne principle, 10.82–84
   history, 91.179–180, 182, 184
   holography, 91.282
   microanalysis See Laser microanalysis
   nonlinear optical effects, 10.84–87
   projections systems, 91.253
   pulsed data, 10.58, 148, 197
   quantum-well laser, 91.187
   radar, 85.66, 68
   ruby laser, 9.247, 250–253; 10.197
   second harmonic generation, 10.84, 85–87
   time-of-flight mass spectrometer, 10.81
   trace element analysis, 10.81
   types, 9.247; 10.197
   YAG laser, 9.247, 250
Laser scanners, 8.36
Laser spot, 83.28
Laser tomography microscopy, 14.230–233
Latent image, photographic, 1.182
Lateral chromatic aberration:
   correction, 14.278–281
   monochromatic, 14.288–289
Lateral magnification:
   and distortion, 14.267
   in microscope objective, 14.253–254
Lateral semiconductor quantum devices, Aharonov–Bohm effect-based, 89.106, 142–178
Lateral structures, mesoscopic devices, 91.213–214, 225
Lateral triodes, 83.68
Latex spheres, water layer thickness determination using, 8.81–83
Lattice, relationship to complex numbers, 84.68
Lattice cell, 83.9, 13
Lattice cell defects, 83.67

Lattice codebooks, 97.218–220, 227–230
Lattice filters, 82.134
Lattice fringes, 11.90
Lattice images, electron microscopy, 7.163
Lattice-ordered group, 84.66
Lattice packing, 97.216
Lattice resolution, 12.31, 49
Lattice transforms:
  applications:
    communications networks, 84.66
    dual transportation problem, 84.124
    generalized skeletonizing technique, 84.115
    image complexity measure, 84.120
    minimax algebra properties, 84.90
    operations research, 84.67
    scheduling problem, 84.94
  defined, 84.71
  image processing, 84.61–127
Lattice vector quantization, 97.194
Law of contradiction, 89.260
Layer-doubling method, 86.212, 216
Lead hydroxide staining, 2.280, 326
Learning:
  quantum neural computing, 94.285–293
  rough set theory, 94.184, 191
Learning from Examples Based on Rough Sets *See* LERS system
Least-squares fit method (LSFM), 13.207
Least upper bound, 84.68
Least-weight path problem, 90.41–42
LED *See* Light-emitting diode
LEED *See* Low-energy electron diffraction
LEED-AES, 11.65
LEEM *See* Low-energy electron microscopy
LEERM *See* Low-energy electron reflection microscopy
Legendre functions, 8.224
Leibnitz's rule, 95.283
Leitz:
  interference microscope, 1.68, 80
  intravital microscope, 6.6
  MMS RT microscope, 5.31, 32
  UMK objective, 6.15

UM objective, 6.15, 16, 17
Lennard–Jones potential, 98.9–10, 11
Lens-coupling fluid interface, 11.156
Lenses; *See also* Microscope objective
  aberrations *See* Aberrations
  cathode, 86.240, 246, 251, 278
  condenser-objective, 10.234–236
  cryolens, 10.39–241
  electromagnetic, 2.5; 10.234–236, 234–241; 91.261–262, 264–265; 93.187–188
  aberration reduction, 8.246–247, 256
  assessment, 5.233, 234
  axial field computation, 13.51–54
  canonical aberrations, 97.381–388
  charged-particle wave optics, 97.282–320, 333–336
  condenser-objective, 10.234–236
  cylindrical, 93.175–176
  design, 2.225
  with elliptical defects, 13.54–59
  FEM and FDM on, 13.129, 153
  field measurement, 2.228
  first-order focusing, 13.182–185
  flux density, 5.212, 213, 214
  focusing, 13.113–114, 182–185
  geometry, 13.52
  Glaser function, 5.213
  Glaser model, 5.212, 213, 214
  Glaser–Schiske diffraction integral, 93.195–202
  IEM on, 13.153–154
  improved phase-object approach, 93.190–192
  integration transformation, 97.369–381, 389–392
  lens field, 13.140–145
  limitations, 5.209
  multislice approach, 93.179, 181–182, 187
  objective lens design, 5.211–214
  optical properties, 83.137
  optimization, 13.207–216
  paraxial properties, 93.194–207
  power-series expansions, 97.361–369

principles of design, 5.209, 210
quadrupole, 93.181–185
round symmetric, 93.186–202
second-zone, 10.236–237
single-pole, 10.237–238
spherical aberration, 1.216;
  93.207–215
spherical wave propagation,
  93.194–195
superconducting, 5.210; 10.238–241
thick lens theory, 93.192–194
electron, 11.68
  characteristics, 2.220
  chromatic aberration, 2.189, 193
  condenser-objective, 2.174
  cylindrical, 93.175–176
  design, 2.225
  diffraction, aberration, 2.172
  objective, 2.224
  quadrupole, 93.181–185
  resolution parameter, 2.173
  round symmetric, 93.186–202
  spherical aberration, 2.171
  stability, 2.190
electron microscopy:
  diamagnetic shielding, 14.20–21
  polepiece, 14.20–21
  solenoid, 14.18–19
  trapped-flux, 14.19–20
electrostatic, 2.5; 10.241–244; 83.62
  aberrations, 8.246
  by synthesis, 13.209–210
  charged-particle wave optics,
    97.320–321
  einzel lens, 96.139
  with elliptical defects, 13.55–57
  FEM and FDM on, 13.129, 153
  first-order focusing, 13.181, 182–185
  history and development, 10.242–244
  IEM on, 13.153–154
  immersion, 10.244
  integration transformation,
    97.396–403
  Japanese contributions, 96.685–688,
    705–707

lens field of, 13.137–140
multislice approach, 93.177–189,
  182–184, 187–188
objective, 10.242–244
optimization, 13.207–216
patent, 96.136
spherical aberration, 1.129
foil, 10.232
high resolution, 10.256–257
image formation, 93.174–176
immersion, 10.244, 245; 83.222,
  228
low voltage SEM, 83.222
multislice approach, 93.173–216
Mulvey's projector, 10.239
objective, double, 10.258, 259
octopole lenses, 1.235, 240, 256
optical, 93.174–176, 185–186
pancake, 10.237, 238; 13.140–141
projector, mirror EM, 4.214
propagation:
  basic equations, 93.175–178
  improved equations, 93.202–207
  light, quadratic index media,
    93.185–186
quadrupole, 86.270, 278
round, 86.239, 249, 254, 260, 278
single pole, 10.237–238
"snorkel," 10.238
superconducting, 10.238–241
  design trends, 14.24–25
  microscopes with, 14.23–24
thick, chromatic effect of, 14.316–317
thin film *See* Thin film lens elements
thin helical:
  focal properties, 14.58–62
  introduction to, 14.58–66
  spherical aberration, 14.62–66
Lens fields, 13.136–145
  electrostatic lens, 13.137–140
  magnetic lens, 13.140–145
Lens method measurement of brightness,
  8.174–177, 182–183
Lens transducer, 11.154
Lenticular screen, 91.254

Le Poole, Jan:
  commercial collaboration, 96.278–279, 287, 289
  metallic and semiconductor materials science, 96.289–292
  microscope development, 96.273–275, 284, 288, 395, 433–435
  solid catalysts, 96.294
  vapor-deposited metal films, 96.293
LERS system, 94.184–189
  All Global Coverings, 94.191–192
  All Rules, 94.192–193
  LEM1 option, 94.188–190, 191
  LEM2 option, 94.190–191
  real-world applications, 94.193–194
  Single Global Covering, 94.188–190, 191
  Single Local Covering, 94.190–191
Lesions:
  detection, 84.338, 341
  focal, 84.338
  SNR, 84.339–340, 342
Leucine aminopeptidase, structure, 7.331, 332
Leucocyte, KFA phase contrast microscopy, 6.77
Leukemia viruses, SiO replication, 8.85
Leukocytes, 5.45, 46
  measurement, 5.69, 70, 71
Levialdi's parallel-shrink algorithm, 90.390–391, 415, 425
Levinson algorithm, 84.303
LFE *See* Local field effect
Lichte focus, 94.219–220
Lie algebra, 84.262, 268; 91.2–3
Lie bracket, 84.139, 84.150, 84.186 ff.
Lie transformation groups (LTG/NP)
  invariant integral transforms, 84.142–157
  invariant pattern representation and, 84.131–192
  visual perception, 84.137–142
Lifshitz "random field approach," 87.53–59
Light-emitting diode (LED), 87.246; 91.179, 182, 246

electronic 3-D system, 91.254
optical symbolic substitution, 89.81–82, 84, 86
Light intensity:
  defined, 10.105, 107
  measurement, 4.391–396
    data processing, 4.396–413
Light optical model, electron beam brightness, 8.185–186
Light-optical reconstruction, electron off-axis holography, 89.5–6, 10–12, 21–24, 33–34
Light pen, 91.235, 244
Light pipe illumination, 6.26–28, 36
  with clad rod, 6.26, 27
  with hypodermic needle, 6.27, 28
Light sensor irradiation, 5.100, 101
Light source:
  beam deflector assembly, 6.23, 25
  for closed circuit television, 6.24
  high-pressure short arc, 6.22
  hypodermic needle type, 6.27, 28
  Köhler system, 6.25, 26
  laser, 6.24, 28
  mounting, 6.23, 24, 25
  for photography, 6.24
  pulsed source, 6.22, 24
  pulsed xenon arc, 6.22, 24, 26
  quartz halogen lamp, 6.22
  rotation, 85.198, 200
  source table, 6.23, 24, 25
Light valve system, 91.252–253
Light waves, principles, 10.104–109
Likelihood function, 97.88
Likelihood ratio distortion measure, 82.139
Likeness relations, 89.296–297
Lindhard dielectric function, 82.199
Linear analysis:
  automatic advantages, 5.126
  contrast resolution, 5.127, 128, 129
  linear analyzer, 5.125, 126
  point-line pattern, 5.124
  using television microscope, 5.130–132
  visual distance quantization, 5.124
Linear convolution, 94.2, 4–7, 12, 19

Linear dichroism microscopy, 8.51–52
　apparatus, 8.63–65
　diagrammatic representation, 8.64
　dichroic ratio, 8.65–66
　methods for analysis of, 8.66–72
　principles, 8.63–65
　wavelength selection, 8.65
Linear dynamic system, 85.18–19
Linear equations, 90.30–35
Linear filters, 92.21, 55–62
Linear lattices:
　imaging of in electron microscope, 1.124
　resolution limit of, 1.126, 134, 136, 148
Linear mapping, 84.281, 310–311
Linear media, constitutive relations, 92.135
Linear minimum square error technique (LMMSE), 92.10
Linear response regime, 89.103
Linear scan, 6.296, 297
Linear scan simulation, 8.40
Linear time-invariant system, signal description, 94.322
Linear transport response, 89.109
Line continuation constraint, 97.130
　extended GNC, 97.132–136
　mean field approximation, 97.131–132
　sigmoidal approximation, 97.137–141
Line pattern enhancement:
　iconic maps, 92.53–55
　linear filter, 92.55–62
　top-hat transformation, 92.63–64
　topographical approach, 92.62–63
Line pattern input, optical diffraction analysis, 7.23
Line process, edge detection, 88.307
Line-shape analysis, bandgap narrowing, 82.261, 263, 266
Line spectrum pair (LSP) parameters, 82.145
Linguistics:
　animal and machine behavior, 94.270–271
　LERS used, 94.194
Liouville equation, 89.118
Liouville's theorem, 8.153, 249; 13.176

Lipase, electron microscopy, 6.195–198
Lipids, 5.299, 300; 6.237
　binary solids, 88.183
　fixation, 2.255, 263, 266
　methylene subcells, 88.155
　phase transitions, 88.176
　retention of, 3.221–228
Lipoprotein envelope, 6.237
Lipoproteins, radiation damage, 5.318
Lippmann–Bragg grating, 10.123, 126
Liquid crystals, 6.120
　development, 91.247
　ferroelectric, 91.247–248
　light valve, 91.252–253
　smectic, 91.247
　as twisted nematic crystal, 91.246–247
Liquid helium cryoshrouds, 11.95
Liquid-metal sources, 11.111; 83.32
Liquids:
　probe–sampling charging, 87.102–106
　for refractometry, 1.64, 65
Lissajous figures, 8.34
Lithium, diatomic molecular bond, model, 98.72–74
Lithium niobate, harmonic images, 10.87, 89
Lithium tungsten trioxide, 91.251
LITHO (software), 13.43, 113
Lithography, 83.110, 111
　electron-beam, 83.36, 76, 110, 112
　ion-beam, 83.110, 111
　light, 83.110
　photolithography, 83.36
Liver, structure, 84.322, 333, 336
Living cells:
　DIC microscopy, 6.105
　phase contrast microscopy, 6.50, 59, 65, 74, 77
　Schlieren microscopy, 6.65
　SEM, 83.209
　stereoscopic microscopy, 6.124
　ultraviolet microscopy, 6.120, 121
Living tissue:
　automated tracking system, 6.38, 39
　clamping, 6.38

Living tissue, *continued*
  mechanically coupled tracking system, 6.44, 45
  microcirculation, 6.2, 46
  motion, 6.28, 34, 38
  seeing in, 6.13
LLL-TV *See* Low light level TV technique
Lloyd's algorithm, 82.113
LLVE *See* Low level vision expert
LMMSE *See* Linear minimum square error technique
Local absolute center, 90.85–86, 94, 106
Local anisotropy measure, 92.53–55
Local contrast enhancement:
  adaptive contrast enhancement, 92.34–36
  extremum sharpening, 92.32–34
  inverse contrast ratio mapping, 92.30–31
  local range stretching, 92.27–30
Local covering, 94.156, 185, 190–191
Local decomposition, 84.84
Local defect, 86.274
Local degree of anisotropy, 92.44
Local field effect (LFE), 12.146; 13.126
Local image-template operation, 90.362
Local labeling algorithm, 90.400–415
Local maximum, 88.301, 335; 90.92
Local minimum, 90.92
Local mode analysis, 82.335
Local neighborhood, 84.78
Local range stretching, 92.27–30
Local spatial order, 92.45–48
Local variation coefficient, 92.15
Local work function, mirror electron microscopy, 4.226–227
Lock-in effect, 85.6
Log-area coefficients, 82.145
Logarithmic Hilbert transform, phase retrieval, 93.111–116, 120–121
Logical operations, 85.148
Logical tasks, animal and machine behavior, 94.272
Logic circuit, 91.149
Logic devices, spin-polarized single-electron devices, 89.223–235
  AND and NAND gates, 89.229–231
  digital systems, 89.233–234
  input and output isolation, 89.238–240
  NOT gates, 89.229
  OR gates, 89.231–232
  performance figures, 89.241–243
Logic gates, spin-polarized single electron, 89.229–232
Loglikelihood:
  Box–Jenkins, 84.297–298
  constrained, 84.267, 282, 301
  forward–backward, 84.296, 298–302, 313–314
  maximization, 84.300–302, 313–314
  unconstrained, 84.264–265
Logons, 97.4
Log-space algorithm, 90.405
Lommel functions, 13.265–266
London penetration depth, 87.187–188
London theory of superconductors, 14.7–9; 87.184
Long cutoff wavelength, defined, 99.76
Longitudinal recording media, 87.159–164
  line charge approximation, 87.159
  stray field, 87.159, 161
Long-term prediction, 82.125
Loop, defined, 90.37
Lorentz deflection, 5.284, 285
Lorentz distribution, in Coulomb interaction, 13.220
Lorentz electron microscopy, 3.183–189
Lorentz equations, 13.174, 178; 90.189–190
Lorentz force, 5.239, 245, 246, 253; 12.94, 123, 126; 95.374
Lorentzian filter, phase retrieval, 93.124–126
Lorentz microscopy, 5.241; 98.422–423
  differential phase contrast mode, 98.350–351
  geometrical optics, 98.351–357
  wave optics, 98.357–358

Foucault mode, 98.360–362
Fresnel mode, 98.335–337
  geometrical optics, 98.337–341
  wave optics, 98.341–350
  particle–wave duality of electrons, 98.333–335
  small-angle electron diffraction, 98.358–360
  small scattering angles in, 7.164
Lorentz oscillations, 87.73
Lorentz trajectories, computation of, 8.243–244
Lorentz transformation, 90.149
Low–Balian theorem, 97.23, 28
Low-beam energy scan; See also High-beam energy scan
  charging effects on insulating specimens, 13.86–92
  collection effficiency, 13.89–91
  and secondary electron computation, 13.89
  simulated line scan, 13.86, 91–92
Low-bit-rate video coding, 97.232–252
  algorithm, 97.237–240
Low-delay speech coding, 82.177, 183
Low-dose imaging, transmission electron microscopy, 94.207–213
Low-energy electron diffraction (LEED), 10.305–306; 11.65; 86.212; 90.256; 94.86–87
Low-energy electron microscopy (LEEM), 94.86
Low-energy electron reflection microscopy (LEERM), 10.306
Lower approximation, of a set, 94.157–165
Lower boundary, rough sets, 94.164
Lower substitutional decision, 94.185
Low-frequency noise, avalanche photodiodes, 99.150, 156
Low level vision expert (LLVE), 86.116–117, 120–122, 124, 133–135, 137–138, 157–158, 163
Low light level TV technique (LLL-TV)
  analog mode, 9.14, 15, 16, 17–22, 46, 49, 59–60
  analog storage system, 9.25–26, 53–54
  background, 9.15–16
  backscattering of electrons, 9.46–48
  CCDs and, 9.58
  defined, 9.56–57
  detection quantum efficiency, 9.8–11, 13–14, 49.52, 58–59
  digital signal processing, 9.22–24
  digital storage system, 9.26–29, 54–55, 58
  dimensioning requirements, 9.8–11
  fiber plates and, 9.51–52, 58, 60
  Fourier transformation and, 9.57–58
  image processing, 9.57–58
  microscope, automatic adjustment, 9.58
  normalization/counting mode, 9.14, 15, 23–24, 46, 49, 58, 60–61
  pixel number, 9.57
  radiation sensitive specimens, 9.56
  resolution, 9.10
  scintillators:
    light absorption, 9.48–49
    primary pulse height distribution, 9.46–52
  signal amplification by quantum conversion stages, 9.14–15
  single-electron response, 9.11–14
  single-image storage, 9.52–55
  slow-scan readout, 9.22
  storage system, 8.55–57; 9.25–29
  tandem optics and, 9.52
  target readout, 9.16–24
  theory, 9.8–29
  TV-image intensifier design, 9.29–46
  video amplifier noise, 9.17, 20
Low power examination, 5.10–16
Low-voltage scanning electron microscopy (LVSEM), 83.203–273
  applications, 83.243–244
    biology, 83.242–248, 251, 254–256
    polymers, 83.249
    semiconductors, 83.249–250
    table, 83.243–244
  backscattered electrons, 83.215, 225–226, 251

Low-voltage scanning electron
microscopy (LVSEM), *continued*
    charging, 83.234–245
    coating, metal, 83.219, 233
    contrast, 83.205–206, 209, 213, 235
    cryo-techniques, 83.252–255
    deceleration, 83.228
    detectors, electron, 83.223–226
    difficulties, 83.221, 222
    early use, 83.220, 222
    economics, 83.225–257
    freeze-drying, 83.232
    freeze-fracture/thaw fix, 83.242, 245
    frozen-hydrated SEM, 83.252–256
    future developments, 83.255, 259
    high resolution, 83.228
    high scan speed, 83.234
    history, 83.205, 256–257
    instrumentation, 83.229
    lenses, 83.222
    modifications, instrumental, 83.230
    optimum voltage, 83.240–242, 253
    performance, 83.230–231
    probe size, 13.292–294
    prototype, 83.256, 259
    radiation damage, 83.208–209, 219, 251
    reemergence, 83.222
    resolution, 83.220–241
    secondary electron imaging, 83.216–218
    semiconductors, 83.227–228
    simulations, 83.218–219
    sources, 83.229
    stray fields, effect of, 83.221
    wavelength, 83.221
    z contrast, SE mode, 83.216
LPCH transform, 84.145
LSP parameters *See* Line spectrum pair parameters
LTG/NP *See* Lie transformation groups
Luckasiewicz operator, 89.278, 288
Luminescence:
    crystal structure and, 99.243
    high level (HL), 82.258
    low level (LL), 82.258
    phosphostimulated *See*
        Phosphostimulated luminescence

spectra, 82.202, 216, 217, 219, 223
LWPP *See* Least-weight path problem
Lymph circulation, 6.1
Lymphocytes, 5.46
Lysosomal acid phosphatase, azo dye technique, 6.186
Lysosomal enzymes, fluorogenic probes of, 14.137–140
Lysosomal hydrolases, 6.189
Lysosomes, 6.176, 182–184, 220
Lysozyme, 5.317, 318

M

Mach effect, 2.41, 46, 55, 63, 64, 65, 69; 3.52
    and size dimensions, 2.73
    theory of, 2.70, 71
Machine intelligence
    linguistics, 94.270–271
    logical tasks, 94.272
    recursive behavior, 94.272
    rough set theory, 94.184
    Turing test, 94.266–272
Machine vision, 97.61–78
    Gabor function, 97.61–66
    Gaussian wavelets, 97.61, 63
Machining tolerances, calculation, 86.225–279
Mach–Zender interferometer, 1.68, 80; 10.113, 176; 12.40; 89.11, 146–149; 99.203–205, 219
Macromolecules:
    diffusion, 6.2
    distribution artifacts, 8.114–115
    interaction with surfaces, 8.113–114
    mounting for electron microscopy, 8.107–132
    physico-chemistry in solution, 8.108–110
    structure preservation, 8.114–119, 222–224
    three-dimensional structure, 7.281–377
Macro volume elements (MVE), 85.162–163

INDEX

MAG3D (software), 13.25, 27
Magnesium oxide, energy selected images, 4.355, 358
Magnetic bottle, 12.96
Magnetic deflection field distribution, calculation, 83.138–139
Magnetic deflectors, 13.149–152
Magnetic dipole moment, 87.146–147
Magnetic distortion field, 94.114–119
Magnetic electron microscope, 87.133–190
   applications, 87.188–190
   commercial production, 96.136–137
   contrast formation, 87.133–138
   contrast modeling, 87.137, 157–181
     electrical current detection, 87.180–181
     interdomain boundaries, 87.176–180
     longitudinal recording media, 87.159–164
     magneto-optic recording media, 87.167–169
     periodic charge distributions, 87.157–158
     type-II superconductors, 87.169–176
     vertical recording media, 87.164–167
   development at Japanese universities, 96.245–247, 251–252
   effective-domain model, 87.134–136, 145–147
   ferromagnetic probe, 87.133–134
   geometrical arrangements, 87.136–137
   interdomain boundaries, 87.189
   invention, 96.132–135
   magnetic microstructure, 98.331, 331–332
   magnetostatic free energy, 87.136
   magnetostatic potential, 87.135
   perpendicular anisotropy, 87.189
   "point-probe approximation," 87.136–137
   resolution, 96.134–135
   scanning susceptibility microscopy, 87.183–188
   sensitivity, 87.181–183
Magnetic field intensity, 82.4, 12, 22, 27
Magnetic fields:
   analytical calculation, 13.128–153
   currents for, arrangement of, 13.141
   and electric fields, combined, 13.107–119
   in electron optical systems, 13.3
   homogeneous, 86.245
   image and, 1.152
   inhomogeneous, 86.240, 245
   numerical calculation, 13.153–173
   quantum phenomena, 92.88
   software for, 13.25–44
   stray, 83.221
   superconductors in, 14.6–15
     general properties, 14.6–7
     Ginzberg–Landau theory, 14.9–11
     London theory, 14.7–9
     magnetic flux structures, 14.13–15
     microscope theory, 14.11–13
     thin film lens elements, 14.73–80
       Biot–Savart law, 14.73–75
       multipole expansion, 14.77–80
       one-turn spiral, 14.75–77
   variations, calculus of, 13.211
Magnetic films, structure of, 3.183–190
Magnetic flux, superconductors, 4.248; 14.13–15
Magnetic flux density, 82.4, 14, 16, 17, 28, 72, 85
Magnetic focusing lens, aberration reduction, 8.246–247, 256
Magnetic fringing fields, mirror electron microscopy, 94.97, 119–124
Magnetic immersion lenses, 13.92–107
   MODEL 1, 13.95–97
   MODEL 2, 13.98
   MODEL 3, 13.99–103
   optical properties, 13.104–107
Magnetic materials:
   domain structure, 98.324, 325–327, 388–389
   electron holography, 98.331, 362–363, 422–423
   principles, 98.363–373

Magnetic materials, *continued*
  STEM holography, 98.373–387, 422–423
  fine magnetic particles, 98.408–409
    chromium dioxide particles, 98.415–422
    phase image simulations, 98.409–415
  history, 98.324
  Lorentz microscopy, 98.422–423
    differential phase contrast mode, 98.350–358, 422
    finite difference methods, 13.24–25
    Foucault mode, 98.360–362
    Fresnel mode, 98.335–350, 422
    particle–wave duality, 98.333–335
    small-angle electron diffraction, 98.358–360
  microstructure imaging techniques, 98.328–333
  mirror electron microscopy, 94.97–98
  multilayer structures, 98.400, 402
    cobalt/copper, 98.406–408
    cobalt/palladium, 98.402–406
  production, 98.324–325
  thin films, 98.387–389
    cobalt, 98.389–397
    nickel, 98.397–399, 400–401
Magnetic moments, 87.155–156
Magnetic monopole moment, 87.146–147
Magnetic phase object, one-dimensional structure, 5.250–252
Magnetic shadow microscope, 91.280
Magnetic shielding, high-temperature superconductivity, 14.38
Magnetic structures, GS algorithm, 7.224, 240
Magnetic thin films, 98.387–389
  cobalt, 98.389–397
  nickel, 98.397–399, 400–401
Magnetic vector potential, 82.16, 26, 41
  canonical aberration theory, 91.5
Magnetic wall, 82.39, 87, 92
Magneto-optical techniques, magnetic microstructure, 98.329

Magneto-optic recording media, 87.167–169
Magnetoplasma physics, 92.80, 81
  anisotropic energy bands, 92.85–88
  band structure anisotropy effect, 92.90–93
  chiral–ferrite media, 92.117, 132
    constitutive relations, 92.119
    dispersion relations, 92.126
    dyadic Green's function, 92.123
  chiral media, 92.117, 132
    constitutive relations, 92.118–119
    dispersion relations, 92.125–129
    electric field polarization, 92.130–132
    vector Helmholtz equations, 92.122–125
  conductivity tensor, 92.83–85, 90–93
  ferrite media, 92.117, 132
    assembly process, 92.179–181
    chiral-ferrite media, 92.117, 119, 123, 126, 132
    circuit parameters, 92.181–183
    constitutive relations, 92.119, 120
    finite-element 2-D equations, 92.162–167
    Helmholtz wave equation, 92.159–162
  finite-element method, 92.132–134, 145–157, 183
    propagating structures, 92.94–95
    2-D equations, 92.162–167
  guided wave structures, 92.183–200
  permittivity tensor, 92.114–117
  planar guiding structures, 92.95–132
    anisotropic determinantal equation, 92.112–114
    dyadic Green's functions, 92.103–106, 108, 112
    normal mode field, 92.95–101
    slot surface fields, 92.108–111
    strip surface currents, 92.106–108
    transformation operator matrix, 92.102–103
  propagation constant, variational formula, 92.93

quantum phenomena, 92.88–90
semiconductors, 92.81–83
variational analysis
  complex media, 92.142–145
  gyroelectric–gyromagnetic,
    92.136–138, 139–141
  propagation constant, 92.93
Magnetoresistance, 91.175
  magnetic multilayers, 98.400, 402
Magnetostatic Aharonov–Bohm effect,
    89.146, 149–156, 166
Magnetostatic field, 82.4, 8
Magnetostatic multipole, 85.245–251
  M function, 85.245–250
  symmetry transformation, 85.250–251
  transformation, 85.247–248
Magnetostriction, 89.237; 98.326
Magnet sector, 86.228, 261, 266, 278
Magnification, in microscope objective,
    14.253–254
Magnitude image, 93.239, 254, 320
Magnitudes:
  geometrical, 5.97
  geometric/geometric comparison, 5.96,
    97
  geometric/optical comparison, 5.96, 97
  measurement, 5.95, 96–99
  optical, 5.97–99
  optical/optical comparison, 5.98, 99
Magnus formula, 97.347–349
Majority filter, 92.65–68
Majorization, 91.47
Malfunction coefficients, 94.393
Malfunction estimation, structural
    properties method, 94.325,
    392–394
Malies–Firth Brown micrometer, 3.50, 51
Malter effect, 83.34
Manifold, 84.137, 181
Many-body effects, 8.218–220; 82.216,
    217, 268
Many-body parameter, 82.205, 220
MAP (maximum a posteriori) estimate,
    97.88, 92, 102–103, 104
Mapping, 97.51, 193

automatic digitization, 84.250–254
backward, 85.124
characteristic mapping:
  classical relational calculus,
    89.275–276, 282
  fuzzy relations, 89.286–287
forward, 85.116
image compression, 97.51–52
optical transforms, 7.55–62
predicate-conditioned:
  sets of regions, 84.235
  subcomplexes, 84.236
SIMD mesh-connected computers,
    90.366–368
τ-mapping, 89.330–332, 383
  anti-extensive, 89.350–351
  binary, 89.336, 366
  cascaded, 89.345–348, 363–366,
    381–383
  dual, 89.348–349, 354–358
  extensive, 89.350–351
  gray-scale, 89.336–337, 366–374
  intersection, 89.339–343, 376–377
  overfiltering, 89.351–353
  translation, 89.337–338
  underfiltering, 89.351, 353–354
  union, 89.338–339, 342–343,
    375–376
template to computer architectures,
    84.78
voxel, 85.215
Mapping problem, 87.284–285
Mapping region, 92.36
Marginal posterior mean, cost function,
    97.102
Marker function, 99.48–49
Markov chain, unified r,s-mutual
    information measures, 91.111–
    112
Markov parameters, 90.108, 111
Markov process, 88.7, 45, 47
Markov random fields (MRFs), 88.307
  Gibbs distributions, 97.106–108
  image processing, 97.90, 105
Martin's diameter, 3.40

Marton, Ladislaus:
  electron microscope, 96.67–71, 102, 183
  heavy metal stain, 96.69, 136, 183
Masking function, 99.292–293
Masks:
  accept masks, 86.108–110
  controlled operation, 86.128
  image, 86.130, 132–133
  reject, 86.108–110
  repair, 12.152
Mass:
  determination, 3.80, 81
  quantum mechanics, 90.159–160, 187
Mass analyzers, with transaxial mirrors, 89.469–476
Massively parallel architecture (MPP), 87.273
Massively parallel computers, 90.364
Mass separators, 11.113
Mass spectrometry, 83.3, 35
  time-of-flight, 10.81
  two-plate electrodes, 89.425–430
Mass transfer, 83.42
Matched filtering, optical symbolic substitution, 89.74
Matching, recognition, 88.272–278
  Dempster–Shafer theory of evidence, 88.278
  feature selection, 88.274
  multivalued recognition, 88.274
  remote sensing, 88.274–276
  rough sets, 88.278
  rule-based systems, 88.277
  syntactic classification, 88.275–277
Match-point, refractometry, 1.85
Mathematical model, fuzzy set theory, 89.255–264
Mathematical morphology, 84.63; 86.89, 107, 165; 89.325–389; 90.356; 99.6, 8
  basis algorithms, 89.334–349, 383–384
  binary τ-mappings, 89.336, 366
  general basis algorithm, 89.337–349, 363–365, 370–374

gray-scale τ-mappings, 89.336–337, 366–374
rank order-statistic filters, 89.335
basis representation, 89.331–333
  filtering properties, 89.349–358
  transforming, 89.374–3383
closing, 84.87; 89.329–330, 336, 345, 372–374; 99.9–10
convex function, 99.13–14
dilation, 84.63, 87; 89.328–329, 343–344; 99.8, 16
erosion, 84.63, 87; 89.328–329, 343–344, 374; 99.8, 16
filtering, 89.349–358
  anti-extensive τ-mappings, 89.350–351
  extensive τ-mappings, 89.350–351
  overfiltering, 89.351–354
  self-duality, 89.354–358
  underfiltering, 89.351–354
general basis algorithm, 89.337
gray-scale function mapping, 89.371–374
τ-mapping, 89.337–349
translation-invariant mapping, 89.363–365
gray-scale function mappings, 89.366–374; 99.8–9
history, 84.86
hit or miss transform, 84.87
image enhancement, 92.48–52, 63–64
kernel representation, 89.330–331
limitations, 84.65
Matheron's theorem, 89.330
median filter, 89.335
multiple transformations, 89.337
opening, 84.87; 89.329–330, 336, 345, 372–374; 99.9–10
operations:
  function, 88.202
  set, 88.202
  skeleton, 88.239
scale-dependent morphology, 99.11–15
theory, 89.326–333
transform, as block Toeplitz matrix with

toeplitz blocks, 84.89
translation-invariant set mappings, 89.361–366
watershed transform, 99.46, 47–48
Mathematics; *See also* Algebra
optical tomography microscope, 14.222–229
missing cone and regularized reconstruction, 14.222–225
non-negative-constrained reconstruction, 14.226–229
support-constrained reconstruction, 14.225–226
rough set theory, 94.151–194
Matheron's theory, 89.330
Matrices, 90.4, 11–13; *See also* Minimax algebra; Minimax matrix theory
adjugate matrix, 90.115
coherence study, 7.172
conjugation matrix, 90.4
convolution property, 94.29–35
definite matrix, 90.56–58
direct property matrices, 94.39–40
Fisher information, 90.134
fuzzy relations, 89.288–289
generalized matrix, 90.112–113
general linear dependence, 90.116
group elements, 94.11–12
Hankel matrix, 90.119
matrix power series, 90.80, 84
notation, 90.6
Pauli spin matrix, 90.164
pointwise maximum and product, 84.71
powering, 90.14
p-regularity, 90.44–45
principal permanent matrix, 90.117
projection matrix, 90.83
scattering matrix, 90.228–229
selvage scattering matrix, 90.246
semidirect tensor product matrices, 94.40–41
square matrix, 90.116, 117
system matrix, 90.4, 14, 79
transitive closure matrix, 90.54–55, 79–80

upper-triangular matrix, 90.53–54
Matrix–matrix operation, 85.25–26, 45–46, 63
Matrix power series, 90.80, 84
Matrix scanning, 6.299
Matrix-squaring, 90.47–48
Matrix transforms, group algebra, 94.27–44
Matrix–vector operation, 85.24–26, 45
Mattauch–Herzog spectrometer, 11.143
Matter, information approach, 90.171–174
Max algebra:
conjugation of products, 90.24–25
greatest-weight path problem, 90.41–44, 80, 84
notation, 90.5–6, 16–19
processes, 90.10–13
MAXIAL (software):
axial fields in, 13.42–44
and Wien filter, 13.115
Maxima of the posterior marginals estimate *See* MPM estimate
Maximum:
local, 90.92
rational functions, 90.106–107
Maximum *a posteriori* estimate *See* MAP estimate
Maximum likelihood estimate (MLE), 85.50; 97.104, 145, 149
exact, autoregressive parameters, 84.298–300
explicit, 84.281–282, 284
linear covariance model, 84.282
member of Jordan algebra, 84.284, 301
Maximum pseudolikelihood (MPL) estimate, 97.148
Maximum spatial frequency, electron off-axis holography, 89.24
Maximum value reprojection, 85.139, 146
Maximum value rule, 84.211, 220
Max/min-median filter, 92.40–42
Maxpolynomials, 90.80, 88–98
characteristic maxpolynomial, 90.118
concavity, 90.107
convexity, 90.107

Maxpolynomials, *continued*
  Evolution algorithm, 90.99–102
  generalized matrices, 90.112–113
  inessential terms, 90.102–103
  merging, 90.88–91
  rectification, 90.105–106
  resolution algorithm, 90.99–103
Maxwell equations:
  electromagnetic fields, 82.2, 3, 5, 16, 19, 20, 26, 27, 28, 36, 38, 39, 41, 42
  from vector wave equation, 90.196–198
  information approach, 90.186–187
  vector derivatives, 95.281–282
MBE *See* Molecular beam epitaxy
McCulloch–Pitts model, neurons, 94.276
MCONT (software), 13.41–42
McWhorter model, 87.215
M-dimensional unified r,s-divergence measures, 91.38–40, 75–95
Mean composition, 89.282–283
Mean curvature, 84.174, 191
Mean filter method, orientation analysis, 93.290
Mean opinion score, 82.101
Mean resultant length, orientation analysis, 93.281–282
Mean resultant vector, 93.287, 293, 298
Mean stage-time, 90.58–59, 71
Measurement, 85.24, 41, 49
  area measurement, 5.120
  automatic linear analysis, 5.126–132
  Heisenberg's uncertainty principle, 90.169
  linear analysis, 5.120, 121
  matrix, 85.17, 19, 21, 47, 55
  noise, 85.24, 38, 41, 48, 65, 68
  optimization, 5.120, 121
  point counting, 5.120
  point-line patterns, 5.121–123
  scalar, 85.21–24, 36
  vector, 85.19, 21, 24–25, 36, 47, 55
  visual linear analysis, 5.123–126
Mechanical gyro, 85.4–5
Medial axis transform (MAT), 85.186
Median multistage filter, 92.40

Median plan, 86.228, 234, 266
Medical diagnosis, fuzzy relations, 89.272, 298–300
Medical imaging, 85.79
Meissner effect, 10.239–240; 14.8, 10
Mellin transform, 84.158
Melting points:
  diatomic molecules, 98.13, 14
  four-body problem, 98.4–13
MEM *See* Mirror electron microscopy
Membrane ATPase, 6.190, 192–194
Membrane crystals, negative staining, 8.123
Membrane filter, LVSEM images, 83.249–250
Membrane potential, fluorescent probe of, 14.152–158
Memory:
  cubic, 85.162
  integrated circuits, 91.148, 151
  visual awareness, 94.289–290
Memory storage, single-spin single-electron devices, 89.233, 234
Mensuration methods, 3.33
Menten method, in alkaline phosphatase microscopy, 6.173, 174
Merging, maxpolynomials, 90.88–91
Merit function, 90.257
MERLIN array, 91.289
Merocyanines, fluorescent probe, 14.158
Mesa structure, 91.214
MESFET *See* Metal–semiconductor field-effect transistors
Mesh-connected computers, 90.356–357
  SIMD mesh-connected computers, 90.353–368
Mesoscopic devices, 89.94, 99; 91.213–228
  Aharonov–Bohm effect, 91.215, 218–219
  anomalous Hall effects, 91.224
  magnetic phenomena, 91.223–224
Mesoscopic physics, 91.215, 291–222
Message passing, 85.268–271
Metachromasy of fluorescence, 8.53

Metal coating *See* Coating, metal
Metal conductors, 83.7, 9, 22
Metal–insulator–metal (MIM) system, 94.124
Metal–insulator–semiconductor (MIS) system, mirror electron microscopy, 94.124–144
Metallurgy:
 and image analyzing computer, 4.364–368, 379–380
 photoelectron microscopy, 10.294–300
Metal oxide, structures, 10.68
Metal oxide semiconductor field-effect transistor (MOSFET), 87.218, 226, 230, 245; 89.98; 91.150, 223
Metal oxide silicon, 83.14
Metals:
 diffusion, mirror EM, 4.226
 mirror electron microscopy, 94.95–96
Metal–semiconductor field-effect transistors (MESFETs), 83.70, 71, 73, 75; 86.32, 36; 99.79
Metal–semiconductor–metal (MSM) photodetector, 99.79
Metal–vacuum interfaces, field electron emission theory, 8.208
Meteorology, 85.79
Methacrylate, embedding in, 2.318, 324
Metioscope, 10.226, 228
Metropolis algorithm, 97.119, 120
Metropolitan-Vickers Electrical Company, 91.261
 EM 2, 96.431–433
 EM 3, 96.435, 441
 EM 3A, 96.450–452
 EM 4, 96.442–443
 EM 5, 96.440–441
 microscope production history, 96.430–433
MFPLOT (software), 13.42, 43–44
M function, 85.237
 determination of constraints, 85.252–253
 possible types, 85.238–239
 symmetry group $GSUB\_sub$, 85.238
 transformation $T(g, v)$, 85.238
MHOLZ *See* Minus high-order Laue zones
MIAS *See* Multipoint multimedia conference system
Micaceous pelite, 7.37
Mica gneiss, 7.37
Mica support film, macromolecule mounting, 8.114, 123–124
Michelson interferometer, 8.29; 10.113, 176
 construction, 6.111, 112
 interference polarizer, 6.111, 112, 113, 114
 interference system, 6.112
 optical symbolic substitution, 89.77
 optical system, 6.111
Microchannel plate, BSE detectors, 83.225–226
Microchip, telecommunications, 91.151, 195, 202
Microcinematography, 4.411; 7.73
Microcircuits:
 edge detection, 8.44, 49
 use of mirror EM, 4.226
Microcirculation, 6.1, 2, 10–13
Micrococcus radiodurans, 5.314, 315, 336
Microcrystals, 6.120
Microdensitometer, integrating, 6.135; 10.9
Microdrops, collisions, 98.13–21, 22
Micro-electric fields, 13.76–78
Microemitter arrays, 83.53, 54
Microfabric, 93.220
Microfabric analysis:
 noisy images, 93.277
 orientation analysis, 93.220–326
 photogrammetric equations, 93.222
 pixel resolution, 93.275–276
 quantitative, 93.221–224
 soil and sediment, 93.221–227, 231, 259, 275–276, 281, 318–319
Microfields, semi-conducting surfaces, imaging, 4.248
Micrography:
 apparatus, 7.82–85
 exposure time determination, 7.87

Micrography, *continued*
  recording techniques, procedures, 7.86–89
  stage movement, 7.83
Microgrid, Japanese contributions, 96.775–778
Microincineration:
  electron microscopy, 3.100–146, 150–153
    accuracy of, 3.118–135, 140, 152
    apparatus for, 3.108, 112, 153
    at high temperature, 3.108–112, 131, 150
    at low temperature, 3.102, 112–116, 131, 144, 150
    for light microscopy, 3.99
    sectioning, 3.107, 141–143, 150, 152
    specimen grids, 3.104, 116, 150
    ultrastructure, 3.143, 151
Microkymography, 7.73–99
  applications, 7.94–99
  indirect, 7.92
  picture content, 7.89–94
  principles, 7.75–80
  procedures, 7.86–89
  quantitative evaluation, 7.89–94
Micromachining, 1.272
Micromagnetic theory, 98.324, 325, 327–328
Microminiaturization, 83.3
Microplasmas, 10.61; 87.236
Microprocessor, 10.14; 86.4; 91.151
Microscope objective, 5.17–19, 101; *See also* Lenses
  aberrations, 14.257–270
    astigmatism, 13.260–265
    chromatic, correction of *See* Chromatic aberration
    coma, 14.266–267
    cover slides, 14.301–309
    curvature of field, 14.265–266
    distortion, 14.267
    monochromatic *See* Monochromatic aberration
    secondary spectrum, 14.276–277, 309–320
    sine condition in, 14.258–259
    spherical, 14.258–259
  achromat objective, 14.277–278, 305–309
  apochromat objective, 14.277–278, 295–297
  bright field microscopy, 7.251–252
  condenser objective system, 6.21, 22
  curvature of field, 14.265–266, 294.298
  design of, 14.249–333
    status and methodology of, 14.298–300
  dipping objective, 6.15–17
  electromagnetic lenses, 5.211–214
  electron lenses, 2.174, 224; 5.211–214; 10.234–246, 242–244
  features, 14.329–333
  focal length, 14.251–253
  focal point, 14.251–253
  focus, 11.37
  high-dry objective, 6.14, 16
  high-voltage electron microscopy, 2.174, 224, 229
  ideal image, 14.254–257
  lateral magnification, 14.253–254
  lens, 83.152, 167–168
  long working distance, 14.320–326
  magnification, 14.253–254
  mirror objective, 6.19–21
  protected objectives, 5.19
  reflected fluorescent observation, 14.326–329
  reflection objective, 6.17
  resolution, 14.254–257
  sine condition, 14.258–259
  single-field condenser-objective *See* Single-field condenser objective
  spectral transmission, 5.17, 18, 19
  telecentric optical system, 14.329–332
  telecentric slit objective, 8.43
  water-immersion objective, 6.16, 17
  zoom systems, 3.3, 15
Microscope theory of superconductors, 14.11–13

Microscopy, 12.279
  Abbe's wave theory, 91.276
  annular pupils, 10.17, 18
  aperture, 5.101
  automatic focusing, 8.26
  automation, 5.39, 43, 44, 45, 50–60, 113
  circular pupils, 10.17–18
  coherent/incoherent light, 10.30, 31
  data processing system, 5.103–106
  definition, sharpness of, 5.102
  diffraction electron microscope, 91.278–279
  digital image processing, 14.215–216
  electron microscopy *See* Electron microscopy
  high-resolution electron microscopy *See* High-resolution electron microscopy
  illumination, 5.101
  image contrast, 5.102
  image formation, 8.3–4
  image quality, 5.100
  image spatial frequency, 10.30, 31
  instrument features, 5.38, 39
  interference, 2.1
  ion probe microscopy, 11.101–148
  of irradiated materials, 5.16–38
  irradiation density, 5.100, 101
  laser illumination, 10.81–82
  laser tomography microscopy *See* Laser tomography microscopy
  light sensor irradiation, 5.100–102, 109–111
  living tissue, 1.2, 5, 15, 29
  low-voltage *See* Low-voltage electron microscopy
  magnetic microstructure, 98.330–333
    Lorentz microscopy, 98.333–362, 422–423
    magnetic force microscopy, 98.331
    mirror electron microscopy, 98.332
    photoelectron microscopy, 98.330
    scanning electron microscopy (SEM), 98.331–332
    scanning electron microscopy with polarization analysis (SEMPA), 98.333
    scanning transmission electron microscopy (STEM), 98.335–337, 342, 353, 357, 358, 360
    spin-polarized low-energy electron microscopy (SPLEEM), 98.333
  magnetic shadow microscope, 91.280
  objective *See* Microscope objective
  optical *See* Optical microscopy
  optical system, 5.17–20, 106–109
  photoelectron microscopy *See* Photoelectron microscopy
  polarization, 2.1
  in polarized light, 5.101, 102
  projection shadow microscope, 91.274, 278
  in reflected light, 5.101
  remote control, 5.20–38
  scanning electron microscopy *See* Scanning electron microscopy
  scanning mechanisms, 8.35–37
  scanning stage, 5.106
  signal display, 5.111–113
  specimen preparation, 5.102, 103
  specimen staining, 5.102
  spectral band width, 5.100
  spin-polarized scanning tunneling microscope, 89.225, 236–237
  substrate examined:
    analytical description, 8.7–10
    qualitative description, 8.5–7
  with superconducting lenses, 14.23–24
  system layout, 5.103, 104
  three-dimensional, 14.214–215
  transmission electron microscopy *See* Transmission electron microscopy
  in transmitted light, 5.101
  types of, 83.203
Microspectrofluorometer, 14.128–130
Microstar system, 6.6
Microstrip, antenna, radiator, 82.1–49
  electric field inside, 82.4–10, 13–15, 27
  Green's function of, 82.3–5, 12
  printed patch, 82.1–49

Microstructure, 93.220
Microtasking, Cray computers, 85.286–292
Microtips, charged, 99.229–235
Microtome *See* Ultramicrotomy
Microtransforms, production, 7.53, 55
Microtriodes, 83.6, 35, 67, 68
Microtubules:
  information transfer, 94.292, 306–307
  three-dimensional structure, 7.345–347
Microvascular organization, 6.2
MICRO-VIDEOMAT, 5.130, 132, 137, 142, 144, 156, 157
Microvilli, 6.77
Microwave beam tubes, 83.35
Microwave conductivity tensor, 92.83–86
Microwave power amplifier tubes, 83.77
Microwave technology:
  history, 91.181–182
  radioastronomy, 91.289–290
  radio relay, 91.193–194
Miller empirical formula, 99.120, 121, 123–124
Mills Cross telescope, 91.287
Mills sample holder, development, 96.50
MIMD architecture *See* Multiple instruction stream multiple data stream architecture
MIM system *See* Metal–insulator–metal (MIM) system
Min algebra, notation, 90.16–19
Mind; *See also* Brain
  consciousness, 94.261–264
  emergent behavior, 94.262
  mind–body problem, 94.261–262
  Vedic cognitive science, 94.264–265, 267
Minerals, 5.99
  image analyzing computer and, 4.380
  microfabric, orientation analysis, 93.226
  orientation analysis, 93.315, 317
Minimal representation, translation-invariant set mapping, 89.360–362
Minimax algebra, 84.66
  Chebyshev approximations, 90.34–35
  components, 90.50
  connectivity, 90.45–51
  critical events, 90.16–26
  discrete events, 90.2–16
  efficient rational algebra, 90.99–119
  equivalent linear programming criterion, 84.102
  infinite processes, 90.75–84
  linear algebra and, 84.71
  linear dependence, 90.36
  maxpolynomials, 90.88–98
  notation, 90.5–6, 16–19
  path problems, 90.36–45
  properties, 84.90
  properties mapped to image algebra, 84.90
    alternating t-t star products, 84.96
    conjugacy, 84.93
    /-defined and /-undefined products, 84.98
    homomorphisms, classification of right linear, 84.92
    linear dependence, 84.103
    linear independence, 84.104
    scalar multiplication, 84.91
  rank:
    column, row O-astic, 84.105
    dual, 84.106
    existence and relation to SLI, 84.106
  scheduling, 90.26–36
  steady state, 90.58–74
  strongly linear independent, 84.103
  systems of equations, solutions, 84.98
    Boolean equations, 84.100
    existence and uniqueness, 84.101
  templates:
    adjugate, 84.107
    based on set, 84.104
    definite, 84.108
    elementary, 84.110
    equivalent, 84.110
    identity, 84.92
    increasing, 84.106
    inverse, 84.109

invertible, 84.109
metric, 84.109
permanent, 84.106
Minimax matrix theory, 84.67
Minimization, truth-table, 89.66–68
Minimum:
   local minimum, 90.92
   rational functions, 90.106–107
Minimum spanning tree, 86.147
Minimum transform:
   minimax algebra, 90.114
   square matrix, 90.116, 117
Minimum uncertainty product, 90.169
Minkowski addition and subtraction, 84.63, 87; 89.328
Minkowski–Bouligand dimension, 88.206–209
Minkowski cover, 88.207, 218
Minkowski inequality, 91.46, 64, 101, 106
Minterms, 89.66–68
Minus high-order Laue zones, 90.300
Mirror-bank energy analyzers, 89.391–478
   charged particle focusing and energy separation, 89.399–410
   charged particle trajectory equations, 89.393–399
   multicascade, 89.430–433, 469–472
   static mass analyzer, 89.472–476
   transaxial mirrors, 89.441–476
   two-cascade, 89.430–432
   two-plate electrodes:
      separated by direct slits, 89.410–432
      transaxial mirrors, 89.441–476
Mirror-based electron analyzers, two-cascade, 89.430–432
Mirror electron microscopy (MEM), 4.161–260; 10.229; 12.96, 97
   applications, 4.226–227; 94.95–98
   without beam deflection, 4.238, 239
   contrast, 4.167–207
      magnetic, 4.258–259
   crystals, 94.96–97
   description, 4.207–217
   history, 94.85–87, 144
   image formation, 94.87–95
      rock salt and, 4.204, 252, 253
   instrument, 94.83–85
   magnetic domain patterns, 4.226
   with magnetic prism, 4.238, 240, 241
   practice, 4.237–249
   projector lens, 4.214
   quantitative contrast technique, 94.98–108, 98–144
      electron distortion field, 94.109–114
      magnetic distortion field, 94.114–119
      magnetic fringing field, 94.119–124
      metal-insulator-conductor samples, 94.124–144
   results, 4.218–226
   scanning mirror electon microscopy, 4.227–232
   shadow imaging, 94.87–95
   shadow projection mirror electron microscopy (MEM), 4.249–259
   specimen perturbation, 4.233–237
   superconductors, 94.98
Mirror prism, for energy analysis, 4.302–318, 342–345
Mirrors:
   charged particle focusing and energy separation, 89.399–410
      transaxial mirrors, 89.433–441
   charged particle trajectory equations, 89.393–399
   electron-optical parameters, 89.393–399, 442–443
   four-electrode mirror, 89.421–425
   mirror with a "wall," 89.420–425
   three-electrode mirrors, 89.417–420, 458–468
   transaxial, 89.392, 396
      charged particle focusing and energy separation, 89.433–441
   two-electrode mirrors, 89.392, 411–417, 441–458
   two-plate electrodes, 89.392
      in mass spectrometer, 89.425–430
      parallel electrode plates, 89.411–420
      wedge-shaped mirrors, 89.430–432

Misalignment, 83.223–224; 86.226, 227, 246
Misfit dislocation, 11.85
MIS system *See* Metal–insulator–semiconductor (MIS) system
Mitochondria, 6.212
  chain enzymes, 6.214, 216
  enzyme distribution, 6.220
  fixation, 2.276, 280, 287
  mineral content, 3.118–123, 135, 144
  oxidation processes, 6.208
  respiration, 6.212, 216
  SiO replication, 8.85, 93
  structural components, 6.216
Mitochondrial ATPase, 6.190
  fixation sensitivity, 6.193
  histochemistry, 6.193
  lead inhibition, 6.193
  location in mitochondria, 6.193
Mitochondrial probes, 14.141–147
Mitochondrial staining, 6.212, 215, 216
  DAB technique, 6.212
  ferricyanide technique, 6.215
  problems, 6.215, 216
Mitotic cell chromosomes isolation, 8.93
Mixed annealing minimization algorithm, 97.122–123, 158
Mixed crystals, semiconductor history, 91.182–183
Mixed time–frequency transforms, signal description, 94.320–321
MKG *See* Microkymography
ML criterion *See* Maximum likelihood criterion
MLE *See* Maximum likelihood estimate
MOCVD, 86.66
Modal analysis, 3.69–77
  in biology and medicine, 3.72
  errors, 3.90–93
  graticules for, 3.73–75
Modal filter, domain segmentation, 93.287–292
Mode:
  analysis *See* Fourier analysis

  physical, 82.63
  spurious, 82.40
Model:
  boundary, 85.96
  constructive, 85.96
  decomposition, 85.96
  imaging, 86.88, 111, 128, 166
  logical, 86.145–146, 165
  mathematical, 85.78
  object, 86.90, 92, 104
  sensor, 86.166
  voxel, 85.100
    binary, 85.101, 199
    gray value, 85.100
    labelled, 85.100
Model-driven parallel architectures, 87.292–296
MODFET *See* Modulation-doped field-effect transistor
Modified forward–backward linear prediction method, 94.370
Modified signed-digit number system (MSD), 89.59–65
  content-addressable memory, 89.70–71
  OSS coding, 89.68–70
Modularity, human visual system, 97.39
Modulated plane wave, 86.193, 195, 199, 204
Modulated-reflectance spectroscopy, 8.44
Modulation, 83.59, 77, 78
Modulation contrast microscope, 8.20
Modulation-doped field-effect transistor (MODFET), 86.54; 89.98
Modulation-doped heterostructure DRAMs, 86.54–58
Modulation transfer function, 11.27; 12.55; 89.32, 34–35; 92.26
Modulator, defined, 99.67
Modus ponens inference, 89.316, 318–320
Moire fringes, 11.82
  crystal aperture scanning transmission electron microscopy, 93.90, 96–97
Molecular beam epitaxy (MBE), semiconductor quantum devices, 89.95, 97

Molecular bonds, diatomic molecules, 98.67–74
Molecular interactions, renormalized, 87.61–65
Molecular scale analysis, van der Waals forces, application, 87.97–100
Molecular structure modification at interfaces, 8.113
Molecules, squeezing *See* Solvation forces
Möllenstedt, Gottfried, 12.4
  biprism, 96.800
  convergent-beam electron diffraction research, 96.585–586, 588–591, 593–595
Moloney leukemia viruses
  contours, 8.91
  freeze-etch replication, 8.93
  "head-to-tail" configuration, 8.90–91, 93
  potassium phospho-tungstate negative staining, 8.91, 94–95
  uranyl acetate staining-critical point-drying preparation, 8.91–93, 94–95
Molybdenum, 83.36, 37, 38
  tips, 83.58, 63
Moment method, 82.1–3, 10–12, 41
  basis functions, 82.1–3, 12–15, 26–27
  convergence rate, 82.1–3, 15, 20–21, 41–43
  matrix elements, 82.13–14, 43–48
Momentum:
  momentum–energy space, 90.157–158, 185
  position–momentum relation, 90.166–167, 187
Momentum–energy space, information approach, 90.157–158, 185
Monge patch representation, 84.149, 189
Monochromatic aberration, correction of, 14.281–298
  curvature of field, 14.294–298
  image plane, 14.290–294
  lateral, 14.288–289
  spherical, 14.282–288
Monochromator, 6.151, 152

Monocytes, 5.46
Monogenic function, 95.303
Monolayer fibroblasts, SiO replication, 8.85
Monolayer overgrowth, 11.60, 61
Monomode fibers, history, 99.71
Monopole lens field, 12.133
Monoscope, 91.240
Monotone property, 99.4, 20–23, 28–29, 47–48
Monotonicity, 89.277, 278
MONTEC (software), 13.217
Monte Carlo methods, 13.74, 76
  and Coulomb interaction, 13.217–218
  electron scattering simulation, 83.217–218, 241–242
  image regularization, 97.119–120
Morgan algebra, fuzzy set theory, 89.256
Morgera–Cooper coefficient, 84.271
Morozov's discrepancy principle, 97.143
Morphological convolution, 99.44
Morphological cover, 88.218, 230
Morphological filter, image enhancement, 92.49–52, 75
Morphological neural networks, 84.62, 89
Morphological scale-space, 99.55
  fingerprints, 99.4–5, 29–37, 55
  future work, 99.56–57
  limitations, 99.55–56
  multiscale closing-opening, 99.22–29, 53, 55
  multiscale dilation–erosion, 99.16–22, 53, 55
MOSFET *See* Metal oxide semiconductor field-effect transistor
Motion *See* Movement
Motion analysis, vision systems, 97.72–74
Motion generator, nonlinear, 3.18
Motor control circuit, 1.20, 36
Motor end-plate, 6.188, 197, 198, 199
  cholinesterase distribution, 6.205
  mammalian, 6.204, 206
  synaptic region, 6.204
Mott formula, 86.219, 221

Mott scattering, spacetime algebra, 95.314–315
Mott's critical concentration, 82.198, 200, 223, 259
Mouse:
  heart muscle, 6.215, 216
  hypothalmus neurone, 6.182
  kidney, 6.210, 217
  L-cells SiO replication, 8.85, 87, 88–91
Movement:
  analysis by microkymography, 7.92
  bacteria, tracking microscope and, 7.1, 5
  determination using microscopes, 7.73
  direct registration, 7.74
  of electron image, 1.150
  quantitative analysis by microkymography, 7.92
  of specimen in electron microscope, 1.120, 132, 150
Move objective lens (MOL), 83.156–159
Moving average (MA) systems, signal description, 94.322
Moving slit diaphragm, recording with, 7.79–80
MPL *See* Maximum pseudolikelihood estimate
MPM (maxima of the posterior marginals) estimate, 97.88, 102
MPP *See* Massively parallel architecture
MQW/SL photodiodes *See* Multiquantum well/superlattice photodiodes
MRF *See* Markov random fields
MSD *See* Modified signed-digit number system
MSETUP (software), 13.31–38
MSM photodetector *See* Metal–semiconductor–metal photodetector
MSOLVE (software), 13.31, 38–39
MTF *See* Modulation transfer function
MTRAJ (software), 13.39–41
Multicascade energy analyzers, 89.430–432, 469–472
Multidimensional Cooley–Tukey algorithm, 93.28–30

Multidimensional parameters, Fisher information, 90.133–135
Multifocus lenses, 12.196
Multifunctionality, semiconductor quantum devices, 89.164
Multigrid convergence, 82.332, 338, 346
Multigrid techniques, 82.328–329
  algorithm, 82.329–347, 364
  analog computers, 82.364
  mixed problems, 82.359–360
  parallel computing, 82.364
  software, 82.365
  Stokes' equation, 82.359
  variational approach, 82.337
Multilayer coil, in magnetic lens, 13.140–141
Multilayered perceptron, 87.7 ff.
Multilayered superconductors, critical current density in, 14.36–38
Multilayer structures, magnetic materials, 98.400, 402
  cobalt/copper, 98.406–408
  cobalt/palladium, 98.402–406
Multimode fibers, history, 99.70–71
Multioutput filter, 94.385
Multiplanar reformatting, 85.133
Multiplanar reprojection, 85.133
Multiple independent collision approach, Coulomb interaction, 13.218
Multiple instruction stream multiple data stream (MIMD) architecture, 85.260–261; 87.262, 269, 272, 280, 294–295
Multiple-slit diaphragm, step microkymography, 7.77
Multiple transformations, mathematical morphology, 89.377
Multiple value fixed radix number system, 89.59
Multiplexed correlator, optical symbolic substitution, 89.82–84
Multiplication:
  linear-time rational calculation, 90.106
  maxpolynomials, 90.89–91
Multiplication shot noise, avalanche

INDEX

photodiodes, 99.151
Multiply connected, eddy current region, 82.31, 35, 36
Multipoint multimedia conference system, 91.205
Multipole, 85.237–239, 251–252
Multipole corrector, 86.278
Multipulse excited linear prediction coding, 82.171
Multiquantum well/superlattice (MQW/SL) photodiodes, 99.94–96
Multiresolution codebooks, 97.202
Multiresolution pyramids, joint space–frequency representation, 97.13–16
Multiscale closing–opening, 99.22–29, 53, 55
Multiscale dilation–erosion, 99.8, 16–22, 53, 55
Multishell lattice codebooks, 97.219
Multislice approach:
  development, 96.51
  electromagnetic lenses, 93.179, 181–182, 187–188
    cylindrical, 93.175–176
    Glaser–Schiske diffraction integral, 93.195–202
    improved phase-object approach, 93.190–192
    quadrupole, 93.175–176, 181–185
    round symmetric, 93.186–202
    spherical aberration, 93.207–215
    spherical wave propagation, 93.194–195
    thick lens theory, 93.192–194
  image formation, 93.174–176
  lens analysis, 93.173–216
    paraxial properties, 93.194–207
  optical, 93.174–176, 185–186
  propagation:
    basic equations, 93.175–178, 186–190
    improved equations, 93.202–207
Multislice simulations, 11.21
Multispectral CT microscope, 14.237–241
  experiments, 14.239–241

principle and theory, 14.237–239
Multispectral processing, orientation analysis, 93.311–319, 321
Multistage one-dimensional filter, 92.39–44
Multiterminal formulas, electron-wave devices, 89.117–118
Multitransputer display engine, 85.166–167
Multivalley ellipsoids, 82.210
Multivariate entropy, unified r,s, 91.95, 102–107
Multiview CT microscope, 14.241–243
Multiwave three-dimensional imaging, with fluorescence, 14.184–189
Mulvey's projector lens, 10.239
Munich electron microscopy, 10.239
Muscle cells, 5.342
Muscle proteins, three-dimensional structure, 7.352–356
Mutual coherence, 7.105–112
Mutual coherence function, 7.108
Mutual intensity, 7.111
  calculation, 7.112
  propagation, 7.112–116
  transfer, 7.117–119
Mutual phase axis, 94.232
MVE See Macro volume elements
Myelin figures, 2.270
Myeloperoxidase, 6.218
Myoglobin, 6.211
Myoneural junction, 6.204, 206
Myosin:
  actin complex, reconstruction, 7.353
  electron micrographs, 7.352
Myosin ATPase, 6.190
  histochemistry, 6.193
  lead inhibition, 6.193
  localization, 6.193

N

NAG, 82.367
Naive labeling algorithm, 90.398–400

NAND gate, 89.229–231
Nanostructure electronic devices, 89.94, 99
Nanotips:
  applications, 95.64, 112, 149
    atomic beam splitter, 95.145–149
    atomic resolution under FEM, 95.112–115
    Fresnel projection microscopy, 95.124–132, 134, 136–140, 142–144
    local cooling, 95.118, 121–124
    local heating, 95.118–121
    microguns, 95.150
    monochromatic electron beam, 95.115, 118
  beam opening angle, 95.97–98
    diffraction through a tunnel barrier, 95.106–107
    geometric effect, 95.105–106
  confinement of field emitting area, 95.84–96
  current saturation, 95.110–112
  current-voltage characteristics, 95.99–100, 110–112
  design, 95.64, 81
  experimental setup, 95.82–84
  field emission:
    characteristics, 95.81–82, 104
    stability, 95.98, 107
  in situ field sharpening, 95.87–88
    buildup tips, 95.91–92
    field surface melting mechanism, 95.93–94
    local field enhancement, 95.88–89
    nanoprotrusion growth and formation, 95.94–96
    sharpening in applied field, 95.89–91
  iron nanotips:
    atomic metallic ion emission, 95.147–148
    beam splitting mechanism, 95.148–149
    field emission, 95.145
  localized band structure, 95.107–110
  local work function decrease, 95.87
  total energy distribution, 95.100–103, 107–110
  ultrasharp tips, 95.85–87
Natural coordinate frame, 86.236
Navigation:
  accuracy, 85.8, 15
  equation, 85.4
  error, 85.11–16, 66, 68
  Kalman filtering, 85.53–70
  system, 85.7, 56
NDR, 89.126, 128
Nd–YAG laser, 10.86
Nearest-neighbor condition, 82.138
Nearest-neighbor interpolation, 85.127
Nearest-neighbor techniques, Coulomb interaction, 13.218
Near field radiation, 12.245
Neel domain wall, 87.179–180; 98.327, 388
Negative differential resistance, 89.126, 128
Negative electron affinity (NEA), 83.10, 29
Negative ions, 11.106
Negative staining, 6.227; 8.108, 116, 123–124
  with ammonium molybdate, 6.235, 263, 267
  of bacteria, 6.228
  of biological macromolecules, 7.288–293
  carbon film technique, 6.261–263, 265, 266, 628
  of collagen, 6.233
  of elastin, 6.233
  electron-dense stain, 6.228
  electron micrograph, effect on, 6.233
  interaction with specimen, 6.234
  with lanthanum acetate, 6.235
  of large protein macromolecules, 6.228
  with lithium tungstate, 6.235
  of membranes, 6.228
  mounting method, 6.230, 231
  optical diffraction technique, 6.229
  pH, 6.227, 228
  with phosphotungstate, 6.227, 228

with phosphotungstic acid, 6.227, 228, 234
photographic averaging, 6.229
with potassium phosphotungstate, 6.228, 235, 260
signal-to-noise ratio and, 7.258
with silicon tungstate, 6.235
with sodium phosphotungstate, 6.228, 235, 263
stain properties, 6.235
stain selection, 6.233, 233–235
of subcellular components, 6.228
with tungstoborate, 6.235
unfixed membrane system, effect on, 6.237
with uranyl acetate, 6.235, 236, 263, 267
with uranyl aluminium formate, 6.235
with uranyl formate, 6.235, 263
with uranyl oxalate, 6.235
of viruses, 6.227, 228
    particle preparation, 6.229, 230–233
wetting agent, 6.234
Neglected residual technique, anisotropic media, 92.132–134, 145–157
Neighborhood system, 97.105
Neighbor interaction function, 97.109
Neon, melting point, 98.10
Nerve myelin, 5.331
Nervous tissue, 94.273–276
Netherlands:
    Delft University of Technology
        bacterial flagella studies, 96.280
        Le Poole microscope, 96.273–275, 288
        microscope construction, 96.275, 288
        organization of electron microscopy research, 96.273–274
        research staff, 96.279
        specimen preparation, 96.279–281
        yeast research, 96.271–273, 276, 281, 283–284
    Dutch Society for Electron Microscopy, 96.298–299
    electron microscopy:
        biological specimens, 96.294–297
        cryofixation, 96.296
        growth, 96.281–284, 295–296
        lenses, 96.288–289
        Philips innovations, 96.289
        software, 96.297
        World War II impact on research, 96.276–278
NETLIB, 82.365
Neumann boundaries, in FDM methods, 13.34–35, 36
Neural networks, 86.74, 87.4 ff., 88.278–288; 94.265; *See also* Deconvolution
    artificial, 94.294
    Daugman's neural network, 97.31, 52
    electronic hardware and, 87.38 ff.
    feedback and feedforward networks, 94.265, 276–278
    generalized Boltzmann machine (GBM), 97.150–151
    Hopfield *See* Hopfield neural network
    iterative transformations, 94.278–284
    models, 94.272–284
    morphological, 84.62, 89
    neuro-fuzzy approach, 88.278–288
        back propagation, 88.282–286
        connectionist expert system, 88.281
        fuzzy neurons, 88.279
        Kohonen's algorithm, 88.283
        perception, 88.280–281
        self-organizing network, 88.281
    optical hardware and, 87.40 ff., 87.44
    optimization, 97.88–89
    quantum neural computing, 94.260–310
Neural process, quantum mechanics, 94.263
Neurons, 94.273–275, 276
    cell death, 94.287
    cholinergic, 6.176, 200
    complex neurons, 94.279–281
    receptive field, 97.40–41
    subneuron field, 94.305–309
Neuropsychology:
    binding problem, 94.284–285
    consciousness, 94.261, 263

Neuropsychology, *continued*
  flawed, 94.294
  learning, 94.285–293
Neuroscience:
  developmental, 94.287
  neural network model, 94.272–276
  reductionist approach, 94.262
Neutrality principle, 89.277, 278
Neutron radiography, 99.264
Neutron topography, magnetic microstructure, 98.330–331
Newtonian mechanics, EPI approach, 90.173
Newton–Raphson iteration formula, 94.279
Newton–Raphson maximization, 82.52, 72; 84.300–302, 313–314
Newton's rings, in quality testing, 10.159–160
Newvicon, 9.343
NF decay length, 12.285
NF optical experiment, 12.281
Nickel:
  magnetic thin films, 98.397–399, 400–401
  photoelectron microscopy, 10.308, 309
Nickel-ion film, 5.285
Nikon condensor, 6.22
Nikon interference-phase device, 6.101–103
Nikon-S-Ke microscope in tracking microscope, 7.3
Nipkow disc, 6.136
Nipkow scanner, 6.140; 10.16, 61
Nitrogen, diatomic molecular bond, 98.72–74
NML microscopes:
  discrimination, 8.38
  edge detection by, 8.45
  scanning mechanisms, 8.35
Noble gases, melting point, 98.9–10
Nodes:
  coherent node renumbering, 90.52
  eigen-node, 90.56
  intermediate node, 90.39
  isolated node, 90.50

Node-set, defined, 90.37
Noise:
  additive, 88.309; 92.10–13
  analog optical fiber communication, 99.68
  autocorrelation function of, 88.310
  avalanche photodiodes, 99.88, 94–96, 150–153, 156
    flicker noise, 99.151–153, 154
    low-frequency noise measurements, 99.150, 156
    multiplication shot noise, 99.151
  Barkhausen noise, 87.240
  binary images, 92.66–68, 72–75
  burst, 87.208, 211, 219, 223, 232, 236
  computer image processing and, 4.95–101
  current noise, 87.206
  defined, 99.76
  diffusion, 87.219
  digital image processing, 4.135–145
    detector noise, 4.141–145
    digital step distribution, 4.136–138
    film grain, 4.138–139
    image sensor, 4.87
    quantization noise, 4.140–141
    system noise, 4.141
  discretization, 92.72–75
  electron off-axis holography, 89.31–32
  emission fluctuation, 83.7, 58
    spectrum, 83.61
  excess, 87.206, 209, 211, 234
  filter response to, 88.312, 319
  flicker, 87.218
  Gaussian, 88.309
  generation–recombination, 87.209, 214, 217, 219, 221, 226, 231, 245
  holographic microscopy, 10.159–173
  homogeneous, 88.310
  imaging plate system, 99.259–260
  Johnson, 87.206
  LMMSE, 92.11–13
  multiplication excess noise, 99.84–86
  multiplicative, 92.13–14
  Nyquist noise, 87.206

INDEX

$1/f$, 87.209, 212, 218, 220, 223, 226, 230, 232, 240, 245
  power spectral $1/f$ noise, 90.174–185, 188
  quantum $1/f$, 87.214
  scanning electron microscopy, 10.73–75
  thermal, 87.206; 88.303, 309
  uncorrelated, 88.310
  white noise, 88.310
Noise removal:
  deblurring, 97.155–159
  image enhancement, 97.56–58
Noise shaping, 97.198
Nomarski method, imaging, 6.105; 10.52, 53
Nomarski variable achromatic system, 6.99, 100
Nonadaptive rank-selection filter, 92.66–68
Nonbinary number systems, optical symbolic substitution, 89.59
Noncomforming elements, 82.340
Noncontact scanning force microscopy, 87.49–196
  capabilities, 87.196
  force-versus-distance curves, 87.195
  instrumentation, 87.190–195
    beam deflection system, 87.192
    bimorph piezosensor, 87.192
    capacitance detection system, 87.191–192
    dynamic mode, 87.193–194
    electron tunneling, 87.191
    force-sensing probe, 87.190–191
    interferometer-based system, 87.192–193
    laser-diode feedback system, 87.193
    setup, 87.194–195
  probe–sample interactions, 87.51–129
    capillary action, 87.52–53
    capillary forces, 87.119–127
    ionic forces, 87.102–112
    method, 87.51–53
    patch charge forces, 87.127–129
    solvation forces, 87.112–119

  van der Waals forces *See* van der Waals forces
Noncontact stylus microscopy, 12.242
Noncontaminating support films, 8.110
Noncrystalline specimens phase solutions, 7.264
Nondestructive readout, 86.49, 69
Nonelementary cycles, 90.39
Nonelementary path, 90.39
Nonelementary state functions, 94.333–335
NonHermitian eigensystems, perturbation methods, 90.250–293
Noninstrumentable quantizers, 84.4
Nonisoplanatism, 89.9
Nonlinear acoustic microscopy, 11.160
Nonlinear gray-value transformation, 92.4–6
Nonlinear laser microscopy, 10.84–87
  shot noise, 10.75
Nonlinear local contrast enhancement, 92.27–36
Nonlinear mean filter, 92.18
Nonlinear transformations, 94.379
Nonlocal type 2 phenomenon, 99.188
Nonmaxima suppression, 88.301, 335, 338
Nonobligatory strokes, 84.245
Nonperiodic structures, perturbation method, 90.272–293
Nonvolatile memory, 86.60, 71
NOR gate, 89.231–232
Normal, to surface, 84.173, 189
Normal component:
  of current density, 82.9, 19, 22, 27, 28
  of current vector potential, 82.20, 28, 36
  of electric flux density, 82.56
  of magnetic flux density, 82.8, 15, 22, 26, 29, 32, 36, 70, 81
  of magnetic vector potential, 82.17, 26, 33
Normal equations, 84.289, 291, 297
Normal mode field, 92.95–101
NOT gate, 89.229

132 INDEX

Nottingham effect, 83.13, 16, 63
  defined, 95.118
  electron field emission, in, 8.253, 255
  nanotip application:
    local cooling, 95.118, 121–124
    local heating, 95.118–121
n-Pentylamine, glow discharge in, 8.126
NPL microscope, 8.35, 40
Nuclear pore complex, LVSEM results, 83.242, 246
Nuclear power, 5.1
Nuclear track scanner, 4.362
Nucleation kinetics, 11.93
Nucleic acid bases, EELS, 9.113, 116
Nucleic acid–protein complexes, 8.125, 126
Nucleic acids, 5.329
  aggregation, 7.288; 8.126
  fixation, 2.260, 262, 267, 280
  negative staining, 7.288; 8.124
  proteins:
    interaction with, 8.127
    loss during sample preparation, 8.119
  single stranded, spreading, 8.127
  solid surface adsorption, 8.126
  specific interaction with proteins, 8.127
  spreading adsorption, 8.119, 124–125
  structure conservation, 8.117–119
  support surfaces for electron microscopy, 8.124–127
  three-dimensional structure, 7.282
Nucleohistone DNA arrangement, 8.66
Nucleohistone–dye complexes, 8.59
  fiber pulling from, 8.60
  orientation, 8.60
Nucleoproteides, three-dimensional structure, 7.282
Nucleoproteins, aggregates with proteins and nucleic acids, 7.288
Nucleoticle base, 5.314
Nucleotidase, 6.182, 183
Nucleus, 6.77, 149
  imaging, 93.87–89, 104
Number systems:
  binary number systems

  half adder, 89.233
  OSS, 89.58–59
  spin polarized single-electron logic device, 89.223–235
  modified signed-digit number systems, 89.59–65
  content-addressable memory, 89.70–71
  OSS coding, 89.68–70
  optical symbolic substitution, 89.58–59
  redundant number systems, 89.60
  residue number system, 89.59
  ternary signed-digit number systems, 89.59
Numerical field electron emission calculations, 8.226–228
Numerical instability, 85.21
Numerical modeling, electron optical systems, 13.59–74
Numerical simulations, contour maps, 99.214–216, 227
Numerical stability, 85.21, 24, 28, 32, 34, 36, 45–46, 50
Nylon fibers, phase contrast photographs and, 8.19
Nyquist noise, 87.206

O

OBIC method, 10.61–65, 68–69; 12.151
Object:
  density function, 5.164
  Fourier transform, 5.164, 165, 167
  optical transfer theory, 4.9–23
  phase-advancing, 6.50, 51, 55, 71, 74, 81
  phase-retarding, 6.50, 51, 55, 56, 57, 71, 74, 81, 87
Objective *See* Microscope objective
Object reconstruction, 93.110–112
  computer simulation, 93.127–130, 133–134, 162–166
Object-scanning:
  advantages, 10.72–73
  and beam scanning, 10.71
  interference microscopy, 10.75–81

methods, 10.70
noise and flare, 10.73–75
Object space, 85.86, 117
   partitioning, 85.161
Object wave function, Fourier transform, 7.244
Oblique projection three-dimensional reconstruction, 7.307
Octopole lenses, 1.235, 240, 256
Octree, 85.107, 122
Octupole, 11.139
Off-axis electron gun rays, energy distribution, 8.192–193, 197–199, 200
Off-axis electron holography *See* Electron off-axis holography
Off–Bragg condition, 11.88
Okayama University, atomic imaging, 96.625–626
Okuto–Crowell theory, 99.136, 141, 148
Oligoclase porphyroblast, 7.37
Olivine, 6.276, 281, 293
OMA *See* Optical multichannel analyzer
One-dimensional phase retrieval, 93.110–112, 131
   deconvolution, 93.145–149
One-dimensional processing, 8.1–2
One-dimensional quantum wire, 91.219–222
One-dimensional symmetry, fast Fourier transform, 93.53–55
110 foil, 93.58
Onsager relation, 89.227
Oolites, optical transforms, 7.38
Opacity, 85.139
Open boundary, 84.223
Opening, 88.202, 239
   function, 99.9–10
   mathematical morphology, 89.329–330, 336, 345, 372–374
      gray-scale, 89.372–374
   multiscale closing-opening, 99.22–29
Open screen, 84.224
Open subcomplex, 84.205
Open subset, 84.201, 204

Operationalism, 94.267
Operation tree, 86.100–101
Operator:
   band pass, 86.111
   morphological, 86.107, 109–111
OPL, commercial microscope production, 96.460
Optical absorption, 82.246, 269
   in Ge and Si, 82.254
   theory of, 82.247
Optical activity tensor, 92.135
Optical antenna, 12.245, 272
Optical astronomy:
   phase retrieval, 93.131
   stellar spreckle interferometry, 93.143–144
Optical beam deflection (OBD) analysis, 12.33
Optical bench, 12.38
Optical bench simulation, 4.115–117
Optical contrast, 6.51
Optical convolution square techniques, 93.222–223
Optical data processing, 8.1
   basic arrangement, 8.4
   cross-grating object, 8.4
   image formation, 8.2, 5
   lens arrangement, 8.2
   phase reversal, 8.22
   spatial filter insertion, effects of, 8.4
   wire mesh object, 8.4
Optical density, 8.17
   defined, 1.182
   measurement, 1.183
Optical destaining, 6.121
Optical diffraction analysis, 3.191, 215; 7.17–71
   bench set up, 7.53
   digital analysis, 7.65
   hardware, 7.51–55
   microscope equipment, 7.53–55
   photographic film, 7.58–60
   transparencies, 7.24, 51, 52
   procedures, 7.51–55
   processing, 7.55

Optical diffraction analysis, *continued*
  schematic diagram, 7.21
  spatial analysis, 7.43–51
  spatial filtering in, 7.62–65
Optical diffraction camera, 6.251, 253
  advantages, 6.251
  camera length, 6.253
  limitations, 6.251, 253
  optimum magnification, 6.253
Optical diffraction pattern:
  computer processing, 6.258, 259
  in electron microscope performance, 6.251
  formation, 6.250, 251, 252, 253
  from amorphous electron micrograph, 6.251
  from periodic electron micrograph, 6.251
  noise, 6.251, 254
  of single virus capsids, 6.261
  of viruses, 6.251, 253, 254, 256, 258
Optical diffractograms, 11.26, 40
Optical diffractometer, 6.250
  calibration, 6.255
  for electron micrograph analysis, 6.250, 253, 254
  mask, 6.255, 257, 258
  for noise removal from optical transform, 6.255
  as spatial frequency filtering system, 6.255
Optical emission spectral analysis, 9.299, 316
Optical examination, 5.10
Optical fiber cables, 99.71
Optical fiber communications, 91.199–207; 99.67–70
  advantages, 99.71–72
  analog, 99.68
  avalanche photodiodes (APDs), 99.81
    germanium APDs, 99.89–90
    InP-based, 99.142–146
    low-noise and fast-speed heterojunction APDs, 99.94–96

  SACGM InP/InGaAs APDs, 99.73, 92–94, 99–156
  SAM and SAGM InP/InGaAs APDs, 99.90–92
  silicon APDs, 99.89
  theory, 99.81–88
  critical device parameters extraction, 99.102–120, 153, 155
  dark current, 99.74, 88, 150–153, 154, 156
  disadvantages, 99.72
  error analysis, 99.110–113, 118–119, 155
  history, 2.12, 13; 91.179–180, 182, 199–201; 99.70–71
  metal–semiconductor–metal photodetector, 99.79
  optical, 99.68, 72–73
  photoconductive detector, 99.79–80
  photodetectors, 99.74–89
  photogain, 99.74, 120–150, 155, 156
  phototransistor, 99.80
  p-i-n photodiode, 99.77–78
  p-n photodiode, 99.77
  Schottky-barrier photodiode, 99.78–79
  temperature-dependence, photogain and breakdown voltage, 99.135–150, 156
Optical fibers, 99.67–68, 70–72
Optical filtering, 7.295
Optical gap, 82.245
Optical gyro, 85.5–7
Optical inhomogeneities, 8.6
Optical interconnects, coupling between, 89.210–213
Optical lens theory, 93.174–176, 186–190
Optically transparent network, 91.210
Optical microscope:
  combined E/S lens, 11.146
  image analysis, 9.323–352
    accuracy of digitized image, 9.333–334
    automated, 9.326, 342–346
    automated analysis, 9.350

automated camera systems,
9.346–348
automated focusing of, 9.326, 343,
345
automated mechanical, 9.326, 343
cost, 9.346, 348
Coulter system, 9.351
digitization, 9.329–334
diode-array detectors, 9.326, 329
flow cytometer, 9.329, 351
"flying spot microscope," 9.328, 330
image processing, 9.346
light pen, 9.337
pattern analysis, 9.337–339
photometric errors, 9.334
picture acquisition, 9.326–329
picture analysis, 9.335–342, 345
picture storage, 9.334–335, 345–346
pixels, 9.330–332, 345
programming languages, 9.340–342
quantitative microscopy and,
9.324–325
recording, 9.342
reporting, 9.342
scanning *See* Scanning optical
microscopy
special purpose systems, 9.348–350
specimen preparation, 9.325–326
television camera and, 9.343
SNOM *See* Scanning near-field optical
microscopy
Optical multichannel analyzer (OMA),
9.281, 282–283
Optical phase difference integration for
dry-mass determination, 6.135
Optical potential, 86.184
electron diffraction, 90.216–217,
341–349
Optical probes, 12.267
Optical processors, optical symbolic
substitution, 89.72–87
Optical properties, 82.206, 244; 83.61, 62
of crystals, 1.41
lithography, 83.110
projection, 83.110

ultraviolet proximity printing, 83.110
Optical pulses, 99.68–69
Optical receivers, 99.72–73
Optical reconstruction, 12.37
computers and, 7.256
of electron micrographs, 7.253–256
Optical resolution limits, 10.14–15
Optical rotation, 1.72, 73, 88, 89
measurement, 1.94, 110
in the presence of binefringence, 1.95
Optical rotatory dispersion measurements,
1.96
Optical sectioning, 10.34–49
automatic focus, 10.37, 39–43, 45
enhancement by processing, 10.55
extended focus, 10.37, 43–47
surface profiling, 10.37–39
Optical shadowing, 7.232
Optical symbolic substitution (OSS),
89.54–58
architecture, 89.71–91
with acousto-optic cells, 89.81–82
with diffraction grating, 89.72–73
image processing, 89.87–91
with matched filtering, 89.74
with multiplexed correlator, 89.82–83
with opto-electronic devices,
89.79–80
with phase-only holograms, 89.74,
77–79
with shadow-casting and polarization,
89.84, 86, 87
coding techniques, 89.57–58, 68–70
content-addressable memory, 89.70–71
image processing, 89.87–91
signed-digit arithmetic, 89.59–71
algorithm for OSS rules, 89.61
higher order MSD arithmetic,
89.61–65
MSD OSS rule coding, 89.68–70
optical implementation, 89.70–71
theory, 89.60
truth-table minimization, 89.66–68
Optical thickness, 8.6
gradients, 8.7

Optical thickness, *continued*
  variation, 8.20; 10.16
Optical tomography microscopy,
    14.213–218; *See also* Laser
    tomography microscopy
  confocal scanning, 14.217
  digital processing, conventional,
    14.215–216
  experiments, 14.233–237
  multispectral microscope, 14.237–241
  multiview microscope, 14.241–243
  optical CT microscope, 14.217–218
  optical transfer function, 14.220–222
  optics, 14.218–222
  phase microscope, 14.243–246
  reconstruction mathematics, 14.222–229
  with rotational off-axial aperture,
    14.229–230
  system configuration, 14.229–237
  X-ray CT technique, 14.216
Optical transfer function, in optical
    tomography microscope,
    14.220–222
Optical transfer theory, 4.1–84
  aperture contrast, 4.71–74
  contrast by phase shift, 4.46–67
  history, 4.80–82
  illumination, 4.7–9
  image formation, 4.1–3
  object, 4.9–33
  optical image, formation, 4.23–45
  partially coherent or incoherent
    illumination, 4.75–79
  point resolution, 4.67–71
  symbols and definitions, 4.3–7
  test objects, 4.79–80
Optical transforms:
  analysis, 7.55–62
  basic properties, 7.22
  examples, 7.34–43
  image analysis, 93.222–223
  mapping, 7.55–62
Optical tunnel microscopy (OTM), 12.301
Optics; *See also* Electron optics
  accelerators, 97.337

charged particles *See* Charged-particle
  wave optics
  electron microscopy, 93.174–176, 215,
    216
  imaging, photomicrography, 4.388–389
  multislice approach, 93.173–216
Optimization:
  of compound scanning systems,
    13.215–216
  of electron trajectories, 13.207–216
    analysis, method of, 13.214–216
    by synthesis, 13.208–214
    dynamic and nonlinear programming
      in, 13.214
    variations, calculus of, 13.209–213
  of lenses, 13.215
  neural networks, 97.88–89
Optimization parameter, 86.126–127,
    129–130, 150
Optimized Dirac–Fock–Slater equation,
    86.221
Optimum aperture, illumination, 1.136
Optimum imaging aperture, 1.122
OPTISF (software), 13.104
Optoelectronic devices, optical symbolic
    substitution, 89.79–80
Optoelectronic integrated circuits (OEICs),
    99.79
Optoelectronic switching, 91.209
Orbit, minimax algebra, 90.13–14, 21–23,
    75–76
Ordered pair, defined, 90.37
OR gate, 89.231–232; 90.18–19
Orientation:
  domain segmentation, 93.287–299
  edge detection, 93.220, 231–244, 259,
    320, 333
  optical diffraction analysis and, 7.43
  soil and sediment microfabric,
    93.221–227, 231, 259, 275–276,
    281, 318–319
Orientation analysis:
  applications, 93.300
  with multispectral processing,
    93.311–319

with porosity analysis, 93.301–311
automation, 93.320–322
image acquisition, 93.228–231
image analysis, 93.220–232, 323–326
  algorithms, 93.319–320
  domain segmentation, 93.278–299
  edge detection, 93.231–239
  index of anisotropy, 93.223, 245–246
  intensity gradient operators, 93.246–287
  quantitative parameters, 93.244–246
  image processing, 93.231–239
  image resolution, 93.275–276
  presentation of results, 93.239–244
  axis conventions, 93.279–280
  quantitative analysis, 93.222–224, 244–246
  statistical analysis, 93.278–294
Orthicon tube, 9.40
Orthogonal functions, signal description, 94.319
Orthogonality
  image representation, 97.6–7, 11, 13, 22
  principle, 82.121
Orthogonalization procedure, 82.169
Orthogonal matrix, 85.27, 29, 33
Orthogonal sum, 94.172
Orthogonal transform:
  in Kalman filtering, 85.26–38, 41, 45, 47, 50–52, 62–63
  sinusoidal transforms, 88.5–10
Orthogonal wavelets, 97.13–16
Orthographic projection, 1.48
Osaka University:
  atomic image processing by fast Fourier transform, 96.622–623
  high-voltage electron microscopy, 96.703–704
  Mark series microscope development, 96.263–267
  materials science research:
    stacking faults, 96.619–620
    vacancies, 96.620, 622
Oscillating mirror microscope scanner, 5.51–53

Oscillograph, 91.263–264
Oscilloscope, 91.233–234
OSS *See* Optical symbolic substitution
Outer normal, 82.5, 15, 22, 34, 38
Outgoing wave Green's function, 90.334
Output, single-electron devices, isolation, 89.238–240
Overdoping, 91.144
Overfiltering, mathematical morphology, 89.351, 353–354
Overlapped block matching algorithm, 97.235, 237–252
Over-relaxation methods *See* Successive over-relaxation (SOR) methods
Oxford digital processing system, 10.14
Oxford model, scanning microscope, 10.3–4
Oxidase, 6.209, 210
Oxidation–reduction enzymes:
  DAB technique, 6.208, 209, 210–212
  electron microscopy, 6.209, 210, 214, 215, 216
  ferricyanide technique, 6.208, 214–216
  histochemistry, 6.208
  optical microscopy, 6.209
  tetrazolium technique, 6.208, 212–214
Oxonols, fluorescent probe, 14.152–156
Oxygen, diatomic molecular bond, 98.72

P

Packing density, 83.38
Palladium/copper, magnetic multilayers, 98.402–406
Pancake lenses, 10.237, 238; 13.140–141
Pancreas:
  acinar cell endoplasmic reticulum, 6.188
  histological section, stripe photomicrography, 7.95
Paraffin crystals, 5.310
Paraffins:
  binary solids, 88.183
  crystal structure analysis, 88.162
  methylene subcells, 88.155

Paraffins, *continued*
  phase transitions, 88.176
Parainfluenza virus, 6.237
Parallel architectures, 87.260
Parallel Computing Forum (PCF), 85.297–299
Parallel image processing, 90.353–357
  Abingdon Cross benchmark, 90.383–389, 425
  algorithms, 90.424–425
    Abingdon Cross benchmark, 90.386, 388–389, 425
    fast local labeling algorithm, 90.407, 410–412
    global operations, 90.369–371
    image-template operations, 90.371–382, 417–420
    labeling of binary image components, 90.396–415
    Levialdi's parallel-shrink algorithm, 90.390–396, 415, 425
    local labeling algorithm, 90.400–415
    log-space algorithm, 90.405
    naive labeling algorithm, 90.398–400
    stack-based algorithm, 90.405, 417
  SIMD mesh-connected computers, 90.353–426
Parallelism:
  explicit, 85.265–271
  human visual system, 97.39
  implicit, 85.263–265
  message passing, 85.268–271
  monitor, 85.267–268
Parallelization, 12.94, 111, 113, 115, 126, 197
Parallel processing methodologies, 87.259–297; *See also*
  Application-driven methodologies;
Architecture-driven methodologies:
  algorithmic characteristics, 87.265–267
  changing data requirements, 87.266
  control parallelism, 87.269
  data access patterns, 87.266–267
  data parallelism, 87.269
  evolution with mix of symbolic and
  numeric processing, 87.263–264
  granularity, 87.273–274
  prototypical view, 87.262
  relationships between architectural and algorithmic features, 87.261–273
  research areas, 87.296–297
Parallel programming, 85.259–265, 297–299; *See also* Cray computers
  approaches, 85.261–263
  implicit parallelism, 85.263–265
  Parallel Computing Forum, 85.297–299
Parallel projection method (PPM):
  projection in image recovery, 95.207–208, 211–212, 215
  recovery with inconsistent restraints experiment, 95.235, 237
  numerical performance, 95.240
  results, 95.238–240
  set theoretic formulation, 95.237–238
Parallel random access machines (PRAMs), 87.279
Parallel-shrink operators, 90.390–396, 417
Parameter estimation:
  channel, 90.128–129
  structural properties method, 94.365–376
Parametric image, 84.344
Parasitic aberrations, 14.97–99
  calculation, 86.225–279
Paraxial electron gun model, 8.140–142
Paraxial properties, electron lenses, 93.194–207
Paraxial ray equation, 93.194, 206
Paraxial Schrödinger's equation, multislice method, 93.200–202
Paraxial theory, 99.179
Paraxial trajectory, 86.231, 236
PARCUM system, 85.162–163, 176
P.A.R. scanning microscope, 1.101, 111
Parseval's theorem, 7.222
Partial coherence, 86.175
  effects, representation, 7.135
  electron image formation and, 7.116–140
  electron off-axis holography, 89.25–31

Partial match, 86.155–156
Particle counter, Lark, 3.59, 60
Particle current density, 92.115
Particle (grain) size parameters, 3.39, 40, 41, 82
Particle mensuration, 4.362
Particle modeling, 98.2
   classical molecular forces, 98.3–4, 67–71
   leap-frog method, 98.4–6, 40
   numerical methodology, 98.4–6, 63–64
   quantitative *See* Quantitative particle modeling
Particle physics, and image analyzing computer, 4.381–382
Particle populations:
   area, 5.145
   automatic analysis, 5.152
   classes:
     arithmetic arrangement, 5.150, 151
     class structure, 5.152, 153
     geometric arrangement, 5.150, 151
     lineal–equidistant arrangement, 5.150, 154, 156, 157
     logarithmic–equidistant arrangement, 5.150, 151, 153, 154, 157
     measurement, 5.152–156
     Rosin–Rammler–Sperling distribution equation, 5.152, 159
   comparative images evaluation, 5.148, 149
   comparison, 5.157–159
   cumulative frequency values, 5.152
   evaluation:
     by circle system, 5.153
     by TGZ 3 particle size analyzer, 5.154, 155, 156
     of diamond powder, 5.153, 154, 155
     from sieve, 5.152
   full characterization, 5.150–156
   mean circumference, 5.146
   mean equivalent diameter, 5.146
   mean particle area, 5.145, 146
   mean particle size, 5.147, 148
   number of intersections, 5.146
   number of particles, 5.145
   partial characterization, 5.145–150
   population comparison, 5.157–159
   STEREOM program to compare, 5.157, 158
   subdivision, 5.152
Particle size, 3.35, 3.39, 3.59, 3.60, 3.63, 3.82 et seq.
Particle size analyzer, 3.59, 60, 63, 64
Particles in suspension, quantitative preparation, 8.121–122
Particle–wave duality, 89.99, 110, 118; 98.333–335
Partition, rough definability, 94.167–171
Partition priority coding (PPC), 97.201
PART-OF relations, 86.86, 143, 154–155, 157–164
Paschen plot, 83.85, 86
Pass filters, generating differential equations, 94.383
Patch charge forces, 87.127–129
Path:
   defined, 84.208; 90.38–39
   in graph, 84.198
   minimax algebra, 90.36–45
Path-connected complex, 84.208
Path weight, 90.39–40
Patterning, 14.54–55; 83.36
Pattern matching, 86.147
Pattern recognition, 86.85, 90, 143, 175
Pauli principle:
   eight-dimensional streamlines, 95.369–372, 374
   fermionic field theory, 95.366
   relativistic wavefunctions, 95.366–369
Pauli spin matrix, 90.164
Pauli spinor:
   momentum density, 95.286
   momentum vector field, 95.286
   observables, 95.285–287
   operator action of matrices, 95.284–285
   parameterization, 95.339
   reflected wave, 95.323
   rotors, 95.287–288

Pauli spinor, *continued*
   spherical monogenics, 95.303–305, 380–383
   spin vector, 95.286
Pauli spin states, two-particle
   basis vectors, 95.352
   causal approach comparison, 95.358–360
   nonrelativistic multiparticle observables, 95.355–357
   nonrelativistic singlet state, 95.354–355
   quantum correlator, 95.353–354
PCF *See* Parallel Computing Forum
PCM *See* Pulse code modulation
PCTF *See* Phase-contrast transfer function
PDA *See* Post-deflection amplification
PDLC *See* Polymer-dispersed liquid crystal film
Pearson's divergence, 91.43
Pechan prism, 14.230–232
PEEM *See* Photoemission electron microscopy
P-elementary sets, 94.154
Pencil of rays:
   aberrations, 8.144–147
   asymptotes measurement, 8.147–148
   crossover, 8.140–144, 145–146, 151
   electron gun, 8.139–140
   emittance diagrams, 8.152–153, 157–162
   entrance pupil at the cathode, 8.141, 145–147
   exit pupil, 8.141–142
   Gaussian intensity profile, 8.152
   geometric properties, 8.150
   longitudinal aberration, 8.145
   ray characterization, 8.149–152
   ray gradient in the cathode image, 8.172
   ray patterns, 8.155–157
   scaling factor, 8.145–147
   shadow curves, 8.148–152
   spherical aberration constant, 8.145
   trajectory calculation, 8.157–162
   transverse aberration, 8.145

Penetration, 83.219
Penetron tube, 91.235–236
Penicillium, 6.108; 8.85
Pennsylvania, geologic map, transform, 7.41
Pepperpot method, electron gun geometrical properties determination by, 8.147
Peptidase, gold technique, 6.199
Perception, binding problem, 94.285
Perceptual criteria, 82.154
Percolation, 87.233, 235
Perfect conductors:
   general properties, 14.6–7
   Ginzberg–Landau theory, 14.9–11
   London theory, 14.7–9
   magnetic flux structures, 14.13–15
   microscope theory, 14.11–13
Period-1 DES, 90.110–111
Periodic charge, distributions, 87.157–158
Periodic continuation approximation, 86.191
Periodic specimens, signal-to-noise ratio, 7.258
Periodic structures, perturbation method, 90.250–272
Periodization, finite abelian group, 93.15
Permalloy film, 5.255, 282, 283, 290, 291
Permanent, minimax algebra, 90.114
Permanent magnets, 98.324
Permeability, 82.3, 7, 13, 23, 39, 40, 41, 52, 61, 72
Permeability tensor, 92.135
Permittivity, 82.3, 7, 40, 61, 92
Permittivity tensor, 92.114–117, 135
Peroxidase, 6.204, 212
   DAB technique, 6.210, 211, 220
   horseradish, 6.204, 210, 217–219
Peroxisomes, 6.212, 220
Perturbation methods, electron diffraction:
   nonperiodic structures, 90.272–293
   periodic structures, 90.250–272
Perturbation theory:
   and aberrations, 13.190

and electron trajectories, 13.173–174, 180
inelastic scattering, 86.186
lens geometry, 13.54–55, 58
nondegenerate, 90.252–255
Petrology:
 and image analyzing computer, 4.381
 optical diffraction analysis, 7.35, 52
Phage butyricum, 7.358, 360, 362
Phase, signal description, 94.318
Phase amplifications, electron off-axis holography, 89.19–21, 47
Phase breaking length, 89.100
Phase-carrying holograms, 8.22
Phase coherence length, 89.100
Phase contrast, 2.205; 11.154; 12.28
Phase contrast/bright and dark field microscopy, 6.116, 117
Phase contrast/brightfield/polarization microscopy, 6.119
Phase contrast devices; *See also* Phase contrast microscopy
 amplitude contrast device, 6.81–86
 with metallic phase rings, 6.72
 nonstandard, 6.63
 positive/negative:
  applications, 6.92, 93
  construction, 6.90, 91
  image faithfulness, 6.92
  negative phase ring, 6.92, 94
  optical system, 6.91
  phase ring composition, 6.92
  phase ring properties, 6.92
  polarizer, 6.92
  positive phase ring, 6.92, 94
  sensitivity, 6.92
 sensitive negative, 6.70–77, 87
  advantages, 6.87
  applications, 6.72–77
  imaging properties, 6.71, 72
  KFA, 6.71, 72, 73, 74, 75, 76
  optical properties, 6.71
  soot phase ring, 6.70, 71
  specification, 6.70, 71
 sensitive positive, 6.77–81, 85, 87
  advantages, 6.87
  applications, 6.79, 80, 81
  KFS, 6.77
  phase ring, 6.78
 with soot phase rings, 6.61, 65–86
 soot phase rings:
  amplitude contrast device, 6.81–86
  highly sensitive negative, 6.70–77
  highly sensitive positive, 6.77–81
  KA, 6.82–86
  KFA, 6.71–77, 92
  KFS, 6.77–81
  in positive/negative phase contrast device, 6.92
 standard, 6.63
 variable:
  advantages, 6.88
  with anisotropic phase plates, 6.96–104
  Beyer, 6.88–90
  for biological specimens, 6.87
  with birefringent phase plates, 6.103
  color phase contrast type, 6.95, 96
  combined positive/negative phase contrast, 6.90–95
  for fine structure, 6.87
  Françon-Normarski system, 6.104
  with isotropic phase plates, 6.87–96
  Nikon interference-phase device, 6.100–103
  Nomarski variable achromatic system, 6.99, 100
  Polanret system, 6.96–99
  polarizer, 6.90, 91, 92, 96, 97, 99, 100, 101, 104, 112, 113
  with re-imaging, 6.96
  with Rheinberg condenser annular diaphragm, 6.96
  with variable characteristic phase plates, 6.96
  variable color-amplitude phase system, 6.103, 104
  Varicolor system, 6.95
  Zernike system, 6.95

Phase contrast electron microscopy, use, 5.240, 295
Phase contrast/fluorescence microscopy, 6.117–119
Phase contrast/holographic microscopy, 6.126–128
Phase contrast/interference microscopy, 6.104–116, 105, 105–116
Phase contrast microscopy, 1.67, 74; 14.243–246
   of biological specimens, 5.299, 300
   contrast threshold, 6.62
   defocusing technique, 5.174, 241, 245, 294
   development, 6.129
   with holographic imaging, 6.126–128
   for image improvement, 5.102, 299
   imaging, 6.57–62, 87; 10.25
   in infrared, 6.120, 121
   interpretation, 6.53–67
   microdensitometric graphs, 6.61, 62
   nomenclature, 6.63, 64
   phase contrast imaging, 6.62
   phase plate, 6.52, 53, 55, 56, 57, 58, 59, 60, 62
   phase retardation, 8.22
   phase ring, 6.52, 53, 88
   in polarized light, 6.119, 120
   positive/negative phase contrast change, 6.90
   principles, 6.52, 53
   quarter-wavelength phase plate, 6.53, 56, 57, 63, 90, 96, 97, 98, 99, 101
   quarter-wavelength phase ring, 6.79, 89
   sensitivity, 6.62, 87, 104
   spatial filtering, 8.20
   stereoscopic, 6.121–126
   system, 6.52
   in ultraviolet, 6.120, 121
   uses, 6.128
   with variable image contrast, 6.88
   vector diagrams, 6.54–57
   vector representation, 6.53–57
   wave separation, 6.55, 56
   Zernike nomenclature, 6.63

Phase contrast/polarization microscopy, 6.119, 120
Phase-contrast transfer function (PCTF), 7.297; 94.209–211, 230, 231, 252
Phase detection, electron off-axis holography, 89.5–7
Phase determination, 7.256–264, 273–276
Phase-difference maps, 99.207–208
Phase distribution, electron off-axis holography, 89.18–19, 41
Phase filters, 8.20–22
Phase gratings, 8.20; 12.53
Phase information, interpretation and use, 7.265–272
Phase memory, quantum devices, 89.100, 111
Phase modulation, 1.87, 88; 7.29
Phase object, 6.50, 51; 12.31
Phase-object approximation (POA), 99.174, 175–176, 184–186
Phase phenomena, in liquids, 6.128
Phase plane, 94.347
   generalized phase plane, 94.324–325, 347–365
Phase plate, 6.53, 55, 56, 57, 58, 59, 60, 62, 63, 64; 12.27, 128
   anisotropic, 6.96
   birefringent, 6.103
   development, 10.232
   negative, 6.53, 56, 57, 61
   positive, 6.53, 56, 57
   quarter wavelength, 6.53, 56, 63, 90, 96, 97, 98, 99, 101, 111, 114
   reflection, 6.104
   variable characteristic, 6.96
Phase problem:
   direct methods for solving, 7.218–238
   in electron microscopy, 7.186–279
   indirect methods for solving, 7.193–218
   in X-ray crystallography, 7.189
Phase range, elementary images, 1.133
Phase refinement, 7.273–276
Phase retardation measurements, 1.77, 109
   by photometry, 1.82
   errors of, 1.100, 101

INDEX 143

with polarizing interference microscope, 1.97
Phase retrieval, 87.3 ff.
  by entire functions, 93.109–168
  algorithms, 93.110–112, 131–133
  blind-deconvolution problem, 93.144
  coherent imaging through turbulence, 93.152–166
  computer simulation, 93.127–130, 133–134
  exponential filter, 93.111–112, 116–118
  Fourier series expansion, 93.118–124, 131–133, 140–142
  Hartley transform, 93.140–143
  Hermitian object functions, 93.123–124, 126, 133
  logarithmic Hilbert transform, 93.111–116, 120–121
  Lorentzian filter, 93.124–126
  one-dimensional, 93.110–112, 131, 145–149
  theory, 93.112–131
  two-dimensional, 93.110–112, 131–139, 161–162, 167
  zero location method, 93.112, 118, 167
  zero sheets method, 93.112, 131, 167
Phase ring, 6.52, 60, 61, 63, 88, 116
  dielectric, 6.92, 94
  narrow, 6.88
  in positive/negative phase contrast device, 6.90, 91, 92, 94, 95
  rapid change, 6.90, 91
  soot, 6.66, 67, 68–70, 71, 72, 77, 82, 92, 94
  soot/dielectric, 6.77, 78, 79
Phase-switching interferometer, radioastronomy, 91.286
Phase trajectories, signal processing, 94.324, 353–365
Phase uncertainty, electron off-axis holography, 89.23
Philco, semiconductor history, 91.144

Philips:
  computer control of microscopes, 96.511–512, 514
  electron microscope, 10.258–259
  EM 75, 96.443–444
  EM 100, 96.433–435, 439
  EM 200, 96.457–458
  EM 201, 96.485
  EM 300, 96.478–479, 481, 485
  EM 400, 96.496–499
  PSEM 500, 96.502
  SEM 505, 96.505
  semiconductor history, 91.144, 177
  350 kV model, 10.257, 259
  TWIN lens, 96.498–499, 514
Phonolite, 7.37
Phonons:
  center of zone, 82.263
  energy of, 82.246
  excitation, 86.187
  generation, 83.29
  LA, 82.246
  LO, 82.246
  momentum-conserving, 82.246
  momentum-conserving TA and TO, 82.256
  TA, 82.246, 259, 263
  TO, 82.246, 250, 259, 263
Phosphatase, histochemistry, 6.179, 180
Phospholipids:
  phase transitions, 8.80
  retention of, 3.224–227
Phosphorescence, 14.122
  and delayed fluorescence imaging, 14.191–196
Phosphors, 2.2, 5, 6; 99.242
  cathodeluminescent, 83.81, 82
Phosphor screen, 11.10
Phosphorylase, three-dimensional structure, 7.339–341
Phosphostimulated luminescence (PSL), 99.242
  applications:
    CBED pattern, 99.269–270
    computed radiography, 99.242, 263–265

Phosphostimulated luminescence
  (PSL), *continued*
  electron diffraction patterns,
    99.270–274
  high-resolution electron microscopy,
    99.274–285
  image processing, 99.285–286
  quantitative image analysis,
    99.274–285
  radio luminography, 99.242, 263–265
  RHEED, 99.286–288
  transmission electron microscopy,
    99.262–263, 265–269, 288
  imaging plate, 99.242, 248–250
  erasing, 99.253
  exposure, 99.250–251
  fading, 99.258–259
  granularity, 99.259–262
  reading, 99.251–253
  resolution, 99.257
  sensitivity, 99.265–269
  mechanisms, 99.242–248
Phosphotungstic acid staining, 2.278
Photoacoustic cell, 12.325
Photoacoustic spectroscopy, 10.58
Photobleaching, fluorescence recovery
  after, 14.174–177
Photocathodes, 10.59; 13.59, 65
Photochromic glass filters, 8.17–19
Photochromism, 8.17
Photoconductance, and surface excess
  carrier density, 10.61
Photoconductive detector, 99.79–80
Photocurrent, avalanche photodiodes,
  99.87
Photodetectors, 99.74–75
  device requirements, 99.76–77
  irradiance, 10.105
  metal–semiconductor–metal
    photodetectors, 99.79
  optical generation, charge carriers, 10.59
  performance characteristics, 99.75–76
  semiconductor photodetectors:
    avalanche photodiodes *See* Avalanche
      phototiodes

  metal–semiconductor–metal
    photodetectors, 99.79
  photoconductive detector, 99.79–80
  phototransistors, 99.80
  p-i-n photodiode, 99.77–78
  p-n photodiode, 99.77
  Schottky-barrier photodiodes,
    99.78–79
Photodiodes, 10.59
  avalanche photodiodes *See* Avalanche
    photodiodes
  InP/InGaAs avalanche photodiodes,
    99–156; 99.73, 90–94
  ionization:
    in absorption layer, 99.110–113
    in charge and grading layer,
      99.113–115
  multiquantum well/superlattice
    (MQW/SL) photodiodes, 99.94–96
  photogain, 99.74, 120–150, 155
  p-i-n photodiode, 99.77–78
  p-n photodiode, 99.77
  Schottky-barier photodiode, 99.78–79
Photo-displacement microscope, 10.58
Photoelectric emission, 83.28
Photoelectric measurements:
  image contrast enhancement, 5.102
  isolation of features, 5.103
  light sensor irradiation, 5.100, 101
  problems, 5.100
Photoelectric setting microscopes:
  accuracy of setting, 8.38–39, 41–43
  advantages, 8.26
  alignment, 8.42–43
  applications, 8.25–27
  defocusing effects, 8.42–43, 49
  depth of field, 8.42
  direct display of line offset, 8.40–41
  discrimination, 8.38–39
  dithered-beam metrology, 8.44
  early types, 8.26, 27
  edge detection by, 8.44–45
  fiducial setting, 8.39, 39–40
  focusing, 8.42–43
  functional parts, 8.26

general layout, 8.27–29
line position measurement, 8.39–41
line structure and asymmetry, 8.41–42
oblique illumination effects, 8.42
operating frequency, 8.36
oscillation system, 8.36–37
repeatibility with, 8.38
scanning mechanisms, 8.35–37
sensitivity, 8.43
setting on a line:
   fixed image scanning, by, 8.31–35
   moving line sensing, 8.37
   radiometric balancing, by, 8.30–31
signal-to-noise ratio, 8.38
synthetic scanning, 8.43–44
system selection, 8.46–47
vertical scale movement, 8.43
Photoelectron, 83.223
Photoelectron emission microscopy (PEEM) *See* Photoemission electron microscopy
Photoelectronic techniques, 1.28
Photoelectron microscopy, 10.228
   acceleration process, effects, 10.276–278
   analyzers, methods, 10.319–321
   applications:
     biological sciences, 10.286, 290, 309–319
     physical sciences, 10.294–309
   Balzer's KE-3, 10.275
   contrast mechanisms, 10.282–287
   conversion of TEM, 10.275
   Davisson–Calbick aperture, lens formula, 10.276
   depth of field, 10.288
   diagram, 10.273
   electron energy analyzer, 10.287
   electron optical system, 10.273
   electrostatic point projection microscopy and, 10.324–326
   emission energy, spherical aberration, 10.278–279
   field emission microscope, photoelectron imaging, 10.326
   future trends, 10.282, 326–327
   as image converter, X-ray absorption, 10.323
   image formation, 10.276–278
   image intensifiers, 10.276
   immunophotoelectron microscopy, 10.315–319
   insulators, 10.300–305
   lasers as light sources, 10.275
   lateral resolution, 10.278–282
   magnetic microstructure, 98.330
   material contrast, 10.284–287
   in metallurgy, 10.285, 287, 294–300
   microelectronics, 10.300–305
   one-lens emission system, 10.271
   Oregon model, 10.274, 280–281
   orientation contrast, 10.287
   photoelectron magnification, x-ray images, 10.323
   photoemission, 10.4
   principles, 10.270–272, 276–294
   related techniques, analytical microscopy, 10.319–321
   resolution, calculation, 10.278–282
   schlieren technique, 10.287
   Schottky effect, 10.283
   semiconductors, 10.300–305
   spectromicroscopy, 10.324–326
   stage process, 10.289
   surface physics, 10.305–309
   synchrotron radiation, 10.276
   topographic contrast, 10.282–284
   ultra-high voltage, 10.274
   X-ray microscope, 10.323
   X-ray scanning, 10.321–322
   zinc silicate binary mixtures, 10.300–303
Photoelectron quantum yields, 10.284–286
Photoelectron spectromicroscopy, 10.228, 270
Photoelectron spectroscopy (XPS), 9.114–115; 12.111
Photoemission, scanned surface, 10.322
Photoemission electron microscopy (PEEM), 8.140; 11.69, 70

Photoemitting scanning tunneling, 12.278
Photogain:
  SAGCM InP/InGaAs avalanche photodiodes, 99.74, 120–150, 155
  temperature dependence, 99.146–150
Photogrammetric equations, microfabric analysis, 93.222
Photographic averaging:
  applications:
    electron microscopy, 6.243
    virus electron micrographs, 6.243
  of gold atomic lattice, 6.244
  linear integration apparatus, 6.243, 246
  loss of contrast, 6.247
  optical diffraction negative analysis, 6.249, 250–255
  photographic emulsion, 6.247
  principles, 6.243
  rotational integration apparatus, 6.243, 245
  stroboscopic apparatus, 6.243, 245
  techniques, 6.243
  of two-dimensional lattice, 6.243
Photographic emulsions, 1.181; 8.17, 20
  choice of, 1.201
  electron diffusion in, 1.199
  electron range in, 1.192
  for electrons, 1.180
  granularity, 1.196
  resolution of, 1.198
Photographic negative:
  contrast, 1.85; 6.228
  image analyzing computer, 4.382
Photographic noise, electron off-axis holography, 89.31–32
Photographic plates, 12.54, 77
Photograph scanner, 4.362
Photography, 1.181; 2.151
  densitometric measurements, 1.86
  development time and temperature, 2.157
  electron microscope, 10.252–253
  exposure temperature, 2.159
  fading, 2.160
  image analyzing computers and, 4.382
  latent image, 1.182
Photoinduced transient spectroscopy (PITS), 10.67
Photoionization microscopy, 10.8
Photolithographic mask monitoring, 8.44–45, 49
Photoluminescence, 82.244, 255, 257, 262
  BGN from observed, 82.259
  biological stains, 8.53
  spectrum, 82.260
Photoluminescence excitation absorption (PLE), 82.256
Photomicrography, 4.385–413
  automation, 4.391–413
    data processing, 4.396–413
    light intensity, 4.391–396
    special instruments, 4.402–413
  instruments, 4.387–388
  range, 4.386–387
  stripe pattern in, 7.90–92
  technical details, 4.388–391
Photomultipliers, 2.2, 10, 81, 82, 86, 90, 92, 107, 108; 4.394–395; 7.12
Photomultiplier tubes, 83.223; 99.75
  analysis of, 13.59–74
  dynode stack, 13.67–74
  front end of, 13.59–66
  noise, 4.141–145
  numerical modeling of, 13.59
  schematic of, 13.60
  sectional views of, 13.62–63
Photon energy, 2.3
Photonic switching, 91.209
Photons, 2.1, 2, 7
  virtual, distribution, 87.53–54
Photon tunneling, 12.269
Photoresistors, 4.396, 397–398, 400–401
Photothermal pulse analysis, 12.345
Photothermal radiometry, 12.327
Phototransistor, 99.80
Phrase grating function, 86.190, 194, 198
Phthalocyanine, relative photocurrent, 10.290

Physical information:
  classical electrodynamics, 90.149–153, 186–187
  divergence, 90.144
  erratum and addendum, 97.409–411
  Fisher information, 90.124–139
  general relativity, 90.170–174, 188
  Poisson information equation, 90.139
  power spectral $1/f$ noise, 90.174–185, 188
  principle of extreme physical information, 90.138, 139–147
  quantum mechanics, 90.154–165, 184, 185, 187
  special relativity, 90.147–149, 186
  uncertainty principle, 90.165–169
  zero information, 90.143–144, 164, 185
Physical optics, 86.176, 177, 187, 193
Picht transformation, 13.181
Pierce electrode, 83.77
Pierce geometry, 11.105
Pierra's extrapolated iteration, 95.211–216
Piezoelectric detection, 12.330
Piezoelectric translators, 83.87
Pillars, 83.82, 83
Pinip storage capacitor, 86.27
p-i-n photodiodes, 99.77–78
Pipeline, flexible, 85.173
Pipeline principle, 85.25
Pisarenko method, 94.369–370
Pitch prediction, 82.125
PITS *See* Photoinduced transient spectroscopy
Pivot point, 94.232, 239
Pixels, 83.79, 80, 207
  defined, 84.73; 85.88; 88.297
  digital image, 88.297; 93.228–229
  domain segmentation, 93.288–289
  edge detection, 93.233
  8-connected, 85.188
  18-connected, 85.130, 191
  electron off-axis holography, 89.24, 32
  finite topology, 84.197, 203
  4-connected, 85.188
  numbering system, 93.233, 268, 284
  optical symbolic substitution, 89.55–56
  rectangular, 93.229, 272, 274–275
  6-connected, 85.130, 144, 188
  SNR, 84.342
  space-variant image restoration, 99.309–12
  square pixels, 93.233, 274, 329
  26-connected, 85.130, 144, 190
  types, 99.303
Pixelwise operations, 90.360, 369, 424
Plagioclase, 6.276, 293; 7.37
Planar guiding structures, 92.95–132
  anisotropic determinental equation, 92.112–114
  dyadic Green's functions:
    admittance, 92.108, 112
    impedance, 92.103–106, 108
  normal mode field, 92.95–101
  slot surface fields, 92.108–111
  strip surface currents, 92.106–108
  transformation operator matrix, 92.102–103
Planar integrated circuit, 91.147–148, 195
Planar monomolecular films, electron microscopy, 7.288
Planar resonance scattering, Bloch waves, 90.313–317
Planar SAGCM InP/InGaAs avalanche photodiodes, 99.96–102, 121
Planar transistor, 91.145–146
Planck's constant, 83.11
Plan-guided analysis, 86.125, 132–133
Plant tissues, 6.120, 210, 212, 230
Plant viruses, 6.228, 234
Plasma display panel, 91.245–246
Plasma frequency, 92.81
Plasma probing, holographic, 10.196
Plasma sources, 11.104
Plasma switching system, 91.250
Plasmatron source, 11.106
Plasmon effects, 12.287; 86.186, 187
Platelets, LVSEM images, 83.249–251
Platform system, 85.3, 7–8, 11, 55
Plausibility function, 94.171
PLE *See* Photoluminescence excitation absorption

Pleochroism, 1.72
Plessy Semiconductors, semiconductor history, 91.141, 147, 149, 150–166, 236
Plinth, fuzzy set theory, 89.265–266
PLMTG (software), 82.365
Ploem illumination system, 6.28, 29
Plumbicon, 9.60, 343
p-n junction, 91.143, 147
　mirror electron microscopy, 4.244
　storage capacitors, 86.6.32
p-n-p alloy transistor, 91.144
p-n photodiode, 99.77
p-n-p storage capacitor:
　charge recovery transient, 86.19
　charge removal transient, 86.21
　C–V characteristic, 86.17
POA *See* Phase-object approximation
Pockels effect, 1.90
POCS *See* Projection onto convex sets algorithm
Poincaré phase plane, 94.347, 350–352, 354–355
Point analysis, 3.250–251
Point cloud, 85.99
Point contact rectifier, 91.142, 143
Point-detector scanning; *See also* Scanning electron microscopy
　comparisons, point source, 10.49–70
　differential phase contrast, 10.49–55
　optical system, 10.2
Pointed fiber, 12.246
Point estimates, image recovery problem, 95.158–159
Point group, 93.9–10
　extended Cooley–Tukey fast Fourier transform, 93.47–52
　reduced transform algorithm, 93.31–39
Point light source Fourier transforms, 8.15
Point-line pattern measurement:
　constant magnitude field, 5.121, 122
　test object, 5.122, 123
"Point-probe approximation," 87.136–137
Point-projection imaging, 10.220, 325–326
Point resolution, 4.67–71

Point sets, 84.73; 90.358
Point-source field electron emission model, 8.221–223, 251, 255–256
Point-source scanning, 10.49–70; *See also* Scanning electron microscopy
Point spread function, 12.28; 93.303, 304
Poisson–Boltzmann equation, 87.111
Poissonian counting error, 6.283, 284
Poisson information equation, 90.139
Poisson statistics, limitations set by in SEM, 83.207–208
Polanret system, 6, 96–99; 6.96–99
Polarization, 86.183, 196
Polarization coding, optical symbolic substitution, 89.57, 59, 84–87
Polarization ellipse, 92.131–132
Polarization interference microscopy, 6.114
Polarized fluorescence microscopy, 8.51–52
　analysis with, 8.66–72
　apparatus, 8.53–55
　background intensity measurement, 8.58
　direction of maximum intensity, 8.58
　initial optimization, 8.59
　principles, 8.53–55
Polarized fluorescence photobleaching recovery, 14.180
Polarizing interference microscope, 1.80, 81, 97
Polarizing microscope, 2.11, 17
Polar molecules, hydrophilic surface preparation, use in, 8.121
Polaron, theory of, 92.89
Polepiece lenses in electron microscopy, 14.20–21
Polycrystalline ferromagnetics, 5.240
Polycrystalline film, 5.240, 289
Polyethylene, 5.309, 316, 317
Polyethylene crystals, 5.307
Polygon, planar, 85.103
Polygon corner, 84.245
Polylysine, 8.126
Polymer-dispersed liquid crystal film, flat panel technology, 91.248

Polymerization, wet replication study of, 8.99
Polymers:
  chitosan, 88.137, 138, 165
  electron crystallography, 88.158, 164–168
  electron diffraction structural analyses, 88.133
  Fresnel projection microscopy, 95.132, 134, 136–139
  lattice images, 88.158
  LVSEM, 83.249–252
  structure, 3.203–206; 88.139
Polymorphism, 88.69, 74, 92
Polynomial filter, binary image enhancement, 92.68–70
Polyoma virus, 7.367, 369
Polyomyelitis virus, three-dimensional reconstruction, 7.367
Polypeptides, structure, 7.288
Polyphos condenser, 6.116
Polystyrene:
  beads, flow, 1.28
  particle suspension preparation, use in, 8.121–122
Pores, transforms, 7.63
Porins, crystal structures of, 88.168–169
Porosity analysis, orientation analysis, 93.301–311
Porro prism reflector, 6.140
Position momentum relation, uncertainty principle, 90.166–167, 187–188
Position-sensitive detectors, 8.30, 31
Positive definite matrix, 85.22, 27, 37, 50
Positively-charged support films, 8.121, 129–130
Positive staining, signal-to-noise ratio, 7.258
POSIX, 88.71
Possibility, rough set theory, 94.166
Post-deflection amplification (PDA), 91.236
Posterior density, 97.100–101
Post-fixation, 2.276, 335
Post-irradiation examination, 5.8, 9

electron microprobe, 5.41
electron microscope, 5.40, 41
low power, 5.10–16
microscope, 5.16–38
scanning electron microscope, 5.41
scheme, 5.9
Postprocessing, digital coding, 97.192–193
Potato virus:
  concentrated preparation, 6.263
  interference pattern, 6.270
Potential:
  distribution, electrical microfield, 94.132–135
  eccentricity, 86.202
  electromagnetism, eddy current fields, 82.25–38
  electron diffraction:
    crystal structure, 90.251, 338–341
    full potential mode, 90.245, 248
    optical potential, 90.216–217, 341–349
    truncated potential mode, 90.244–245
  fluctuation, 82.232, 235
Power density, 83.30, 35, 63
Power fluctuation, image resolution and, 7.262
Poweroid structuring functions, 99.14–15, 41, 42–46
Power series:
  expansions:
    of eikonals, 91.16–28, 34
    for Hamiltonian functions, 91.8–13
  matrix power series, 90.80, 84
  minimax algebra, 90.79–84
  scalar power series, 90.80–83]
Power spectral density (PSD), signal description, 94.322–324
Power spectral $1/f$ noise, 90.174–185, 188
Poynting flux, 90.150
Poynting theorem, 82.40
Poynting vector, 82.87, 90
PPM *See* Parallel projection method
PRAMs *See* Parallel random access machines
Predicate, 84.234, 245

Predicate calculus, 86.82, 86, 144, 146, 164
Predicate-conditioned mapping
 subcomplexes, 84.235, 236
Prediction error sequence, 82.122
Prediction vector, 94.369
Predictive coding, image processing,
 97.51, 193
Predictive speech coding *See* Speech,
 coding
P-regularity, 90.44–45
Preprocessing, digital coding, 97.192–193
Prescriptive learning, 94.286–288
Pressure tests, 83.53
Prewitt operator, 92.34; 93.235, 236, 253,
 264, 266, 267, 269, 277, 323
Primal sketch, 97.64
Primal weighting, 90.37, 38
Primary electrons:
 and deflection fields, 13.146
 and EBT, 13.126
 and electrostatic charging, 13.226–227
 interactions, 13.76–77
Primary interference patterns, 8.3
Prime factor algorithm, 93.2
Principal idempotents, 84.275
Principal permanent matrix, 90.117
Principal permanent mean, 90.117
Principal vibration directions, 1.46, 51
Principle of extreme physical information
 (EPI), 90.138, 139–147, 178–179,
 185–190
 classical electrodynamics, 90.153,
  186–187
 general relativity, 90.170, 173, 188
 power spectral $1/f$ noise, 90.176,
  179–180, 183–185, 188
 quantum mechanics, 90.154, 160, 164,
  184, 185, 187
Prior density, 97.88
Prior–DFT estimator, 87.13 ff., 87.31 ff.
Prior information, 97.100
Prismatic scanner, 6.137–142
Probabilistic error, 85.53
Probabilistic sum, fuzzy set theory,
 89.261–262

Probability density function, states of
 information, 97.98–101
Probability of error, unified r,s-divergence
 measures, 91.120–132
Probability law–estimation procedure,
 90.145–146
Probability theory, evidence theory,
 94.171–182
Probes:
 in fluorescence microspectroscopy,
  14.168–171
 cell cations, 14.160–167
 of cell cytoskeleton, 14.150–152
 DNA, hybridization of, 14.172
 and electron microscopy, 14.172–174
 endoplasmic reticulum, 14.147–150
 gene expression, 14.171–172
 for Golgi apparatus, 14.140–141
 latest methods in, 14.168–174
 lysosomal enzymes, 14.137–140
 membrane potential, 14.152–158
 mitochondrial, 14.141–147
 proton indicators, 14.158–160
 formation, 10.229
Procentual composition, 89.283–284
Processing elements, arrays, 87.274, 277
Product forms, extrema, 90.91–93
Profile plan, 86.232
Profile velocity, 1.5, 16, 27
Proflavine (PFA), 8.53
Programmable ROMS (PROMs), 86.71
Programming:
 parallel programming, 85.259–265,
  297–299
 SIMD mesh-connected computers,
  90.357, 369–382
Projected potential, 86.189, 201
Projection:
 finite number, reconstruction from,
  7.327, 328
 orthographic, 85.119
 perspective, 85.114, 119
 symmetry properties, 7.307–310
Projection display, 91.252–253
Projection matrix, 90.83

Projection onto convex sets algorithm
(POCS):
  image recovery, 95.179–181, 183,
    200–201, 238–240
  image restoration with bounded noise,
    95.248
  limitations of convex feasibility problem
    solving:
    countable set theoretic formulation,
      95.202
    inconsistent problems, 95.201–202
    serial structure, 95.201
    slow convergence, 95.201
  practical considerations, 95.232–234
  subgradients, 95.256–257, 259
Projection shadow microscope, 91.274,
  278
PROMs *See* Programmable ROMS
Propagation:
  basic equations, 93.176–177, 186–190
  charged-particle beam, scalar theory,
    97.279–282
  electromagnetic, highly anisotropic
    media, 92.80–200
  Gaussian wavefront, 93.204–207
  image formation, 93.174–176
  improved equations, 93.202–207
  light, in quadratic index media,
    93.185–186
  quadrupole field, 93.181–185
  spacetime algebra, 95.310–311
  spherical wave in lens field, 93.194–195
  spinor potentials, 95.311–312
  spinor theory, 97.330–332
Propagation coefficient, 82.86
Propagation constant, variational formula,
  92.93
Propagation function, 86.187, 188, 189,
  196, 197
Proportionality factor, 94.207
Proteins, 5.305
  aggregate with nucleic acids and
    nucleoproteins, 7.288
  bacteriorhodopsin, 88.168
  complex structure conservation, 8.117

  in crystals and layers, 7.330–332
  fixation, 2.258, 263, 265
  porins, 88.168 ff.
  radiation damage, 5.317, 318
  structure, 7.288
Proton indicators, fluorescent probe,
  14.158–160
Protoplasm, 6.175
Protozoa, 5.314, 336
PSD *See* Power spectral density
Pseudocholinesterase:
  acetylthiocholine technique, 6.200
  function, 6.200
  occurrence, 6.200, 206
Pseudoclassical approximations, validity
  criteria, 5.268–285
Pseudoconcavity, 91.66–69
Pseudoconvexity, 91.47, 66–67
Pseudo-inverse matrix, 85.17
Pseudopotential, 86.221
Pseudoraster, 84.253
Pseudo-stereo system, 91.253–254
PSL *See* Phosphostimulated luminescence
Psychons, 94.306
Psychophysics, vision modeling, 97.35, 39,
  40, 43
Pulsation, 1.4, 6, 27, 30
Pulse code modulation (PCM), 91.196,
  202–204
Pulsed dc electroluminescence, 91.245
Pulsed laser beam, 12.145
Pumice, vesicular structure, 7.37
Pupil (eye):
  area microkymography, 7.97
  function:
    monochromatic system, 10.17
    single lens, 10.16–19
  movement, microkymography and, 7.96
  registration, 7.74
Pupillokymography, infra red, 7.74
Purine bases, DNA stacking orientation,
  8.64
Purple membrane, 8.115
  amplitude information, 7.275
  collapse, 8.116–117

Purple membrane, *continued*
 contour map, 7.209
 CTEM vs. STEM, 7.210
 electron diffraction pattern, 7.204, 206
 electron micrograph, 7.301
 image reconstruction, signal-to-noise
    ratio, 7.215
 micrographs, optical transform
    quadrants, 7.206
 negative staining, 8.123
 projected map, 7.208
 reconstructed image, 7.302
 three-dimensional structure, 7.335–337
 unstained, electron micrograph, 7.302
Pyramidal image model, 92.36–38
Pyrimidine bases, DNA stacking
    orientation, 8.64
Pyroxene, 6.276, 293

Q

QPC *See* Quantum point contact
QR algorithm, 85.33–34, 63
QUADFET *See* Quantum diffraction
    field-effect transistor
Quadratic filter, 92.68
Quadratic index media, light propagation,
    93.185–186
Quadratic structuring functions, 99.15,
    45–46
Quadrature, numerical, 82.340
Quadrature pyramid, joint space-frequency
    representations, 97.19, 20
Quadrupole lenses, 11.114
 aperture aberrations:
    calculation, 1.246
    correction, 1.255
    effect of pole shape, 1.253
    measurement, 1.250
 astigmatic doublet, 1.251
 combinations equivalent to round lenses,
    1.241
 electrostatic quadruplets, 1.242
 electrostatic triplets, 1.244, 260

 mechanical defects of, 1.263
 multislice approach, 93.181–185
Quadrupole-octopole lens, 1.259
Quad tree, 86.147
Quality of lower/upper approximation,
    rough sets, 94.181
Quantified image, fuzzy relations,
    89.289–290
Quantimet image analyzing computer,
    4.361–383
Quantimet microscope, 5.44
Quantitative area scanning:
 effect of instability, 6.284, 285
 proportional drift correction, 6.285
 in pseudo-random pattern, 6.285
Quantitative image analysis, 5.115;
    93.222–224; 99.274–285
Quantitative imaging, in fluorescence,
    14.183–184
Quantitative mirror electron microscopy,
    94.98–144
 electron distortion field, 94.109–114
 magnetic distortion field, 94.114–119
 magnetic fringing field, 94.119–124
 metal–insulator–semiconductor samples,
    94.124–144
Quantitative particle modeling, 98.2
 classical molecular forces, 98.3–4
 crack development, in stressed copper
    plate, 98.21, 23–30, 31–33
 diatomic molecules:
    melting points, 98.13, 14
    molecular bonds, 98.67–74
 drops:
    colliding microdrops of water,
        98.13–21, 22
    liquid drop formation on solid
        surfaces, 98.30, 32–44
 fluid bubbles, 98.44–61
 melting points, 98.6–13
 numerical methodology, 98.4–6
 rapid kinetics, 98.61–67
Quantization, 82.110; 84.2
 adaptive quantization, 82.114
 defined, 97.51, 193

INDEX

digital image processing, 4.135–145
  detector noise, 4.141–145
  digital step distribution, 4.136–138
  film grain, 4.138–139
  quantization noise, 4.140–141
  system noise, 4.141
image compression, 97.51
optimal quantization, 82.113
scalar quantization, 82.110–111, 117; 97.193, 194
  finite state scalar quantization, 97.203
  wavelets, 97.200–201, 202–203
successive approximation quantization
  convergence, 97.211–214
  orientation codebook, 97.214–216
  scalar case, 97.205–206
  vectors, 97.207–211
successive approximation wavelet lattice vector quantization (SA-W-LVQ), 97.191–194, 252
  coding algorithm, 97.220–226
  image coding, 97.226–232
  theory, 97.193–220
  video coding, 97.232–252
uniform, 82.111–113
vector quantization, 97.51, 193
  wavelets, 97.201–202, 203
Quantizers
  equivalent quantizer, 84.13
  exhaustive search vector quantizer, 84.3, 5, 22, 39–45, 52, 57
  lattice vector quantizer, 84.4
  rate distortion data, 84.52–57
  residual quantizers, 84.1–52
    exhaustive search, 84.2, 5, 22, 39–45, 52–57
    reflected, 84.30–37, 45–50
    scalar, 84.14–26
    vector, 84.26–30
  single-stage quantizers
    scalar, 84.6
    vector, 84.9
  tree structured quantizers, 84.4
  trellis-coded scalar quantizers (TCQ), 84.51

Quantum bubble, 89.97
Quantum chips, 89.208–243
Quantum computation, neural computation, 94.298–305
Quantum conductance, 89.113–117
Quantum-confined systems, 89.96
Quantum-coupled architectures, 89.217–243
Quantum dashes, 89.219, 220, 224
Quantum devices:
  defined, 89.98
  reproducibility, 89.222, 244
  semiconductor, 89.93–245
    Aharonov–Bohm effect-based devices, 89.106, 142–178
    connecting on a chip, 89.208–243
    coupling, 89.208–243
    directional couplers, 89.193–199
    electron wave devices, 89.99–120
    granular electronic devices, 89.203–208
    quantum-coupled devices, 89.208–243
    resonant tunneling devices, 89.121–142, 222, 245
    shortcomings, 89.210–222
    spin precession devices, 89.106, 199–203
    transistors, 89.106, 157, 160–165, 167, 189, 244
    T-structure transistors, 89.178–193
  superconductor, 89.98
  tunnel diode, 89.98
Quantum diffraction field-effect transistor (QUADFET), 91.226–227
Quantum dots, 89.97, 219, 220, 224
Quantum efficiency:
  avalanche photodiodes, 99.87
  photodetectors, 99.75
Quantum interference effects, 89.106–110, 120
Quantum interference transistor, 89.106, 157, 160–165, 167, 178, 244
Quantum mechanical coupling, 89.213–217, 221–223

Quantum mechanical effects, mesoscopic devices, 91.213, 214, 216–218
Quantum mechanical tunneling, 89.136–142, 216
Quantum mechanics, 84.262–263, 276; 89.99
  Copenhagen interpretation, 94.297
  information approach, 90.154–165, 184, 185, 187
  neural processes, 94.263
  operationalism, 94.267
  quantum neural computing, 94.260–310
  resonant tunneling devices, space charge effect, 89.130–133
  Vedic cognitive science, 94.264–265, 267
Quantum model, 94.296–298
Quantum neural computing, 94.260, 266, 296–298, 300, 302, 309–310
  binding problem, 94.284–285
  complementarity, 94.295–296
  consciousness, 94.261, 263
  defined, 94.298
  learning, 94.285–293
  mind–body problem, 94.261–262
  neural network models, 94.272–284
  quantum and neural computation, 94.298–305
  structure and information, 94.305–309
  Turing test, 94.265, 266–272
  uncertainty, 94.294–295
Quantum noise, 1.195; 12.43, 47
  electron off-axis holography, 89.31–32
Quantum phenomena, magnetoplasma physics, 92.88–90
Quantum physics:
  complementarity, 94.262–263, 295–296
  consciousness, 94.263
Quantum point contact, 91.219–222, 224
Quantum theory, 14.123
  charged particle wave optics, 97.257–259, 336–339
    aberrations, 97.311–312
    scalar theory, 97.259–322
    spinor theory, 97.258, 322–336

Quantum well, 89.97; 91.213, 215
  coupling, 89.195–199
  double quantum wells:
    electrostatic, 89.157–178
    magnetostatic Aharonov–Bohm effect, 89.149–156
  resonant tunneling devices, 89.121–142
Quantum-well floating-gate DRAMs, 86.49–54
Quantum-well laser, 91.187
Quantum-well memory, 86.49
Quantum wire, 89.97
  Aharonov–Bohm interferometer, 89.172, 178, 180, 186, 199
  one-dimensional, 91.219–222
Quarteoze pelite, 7.37
Quarter wavelength plate, 6.53, 56, 57, 63, 90, 96, 97, 98, 99, 101, 111, 114
Quarter wave phase ring, positive, 6.79, 89
Quartz, 7.37; 87.224
Quartzite, photomicrograph, transforms, 7.65
Quartz oscillator gauge, 11.62
Quasicomplete Gabor transform, 97.34–37
Quasiconvexity, 91.47, 67
Quasi-darkfield microscopy, 6.116
Quasi-dissipative electron transport, 89.118–119
Quasi-elastic scattering, electron diffraction, 90.217–218
Quasi-static memory, 86.71
Quaternion groups:
  index notation, 94.9–10
  matrix representation, 94.13
  matrix transforms, 94.36–39
Quiescent plasma, 11.106
Quotient ring, group algebra, 94.25
Quotient space, 84.202, 225, 272–273, 286
Quotient topology, 84.225

R

Rabbit:
  mesentery, 6.38
  muscle fiber, 6.205

Radar:
    flat panel technology, 91.245
    history, 91.143, 261, 268
    measurement, 85.54, 56, 66
    smoothing of SAR speckle noise, 92.13–14
Radially ordered circulator, z-ordered layers, 98.260–283
Radial stray field, 87.144–146
Radiation biology:
    comparison with electron microscopy data, 5.311, 312, 313
    dosage units, 5.311
    linear energy transfer values, 5.311
Radiation damage, 2.196, 236, 400; 11.51; 83.208–210, 219, 237, 251, 254, 256; 86.30
    beam-induced conductivity, 83.234
    cathodeluminescence (CL), 83.214
    deformation, shrinkage, 83.210, 256
    effects, 83.209, 212, 214, 237
    in electron microscopy, 7.289
    image interpretation and, 7.186
    inelastic electron scattering, 7.263
    living specimens, SEM, 83.209
    measurement, 5.307, 311
    phase determination and, 7.191
    phase information and, 7.256, 273
    phase problems and, 7.257
    reduction, 5.309, 310
    repair, 5.337, 338
    resolution and, 7.192
    semiconductors, 83.228
    specimen structure and, 7.213
Radiation dose, 83.209
Radiation hard devices, 83.70
Radiation-sensitive specimens, damage-free imaging, 9.6–7
Radioactive tracer, image intensifier detection of, 2.39
Radioactivity, 5.2
Radio antennas, 12.245, 248
Radioastronomy, 91.285–290
    band limited constraint, 7.244
    interferometers, 91.286, 287, 289

Radiofrequency echogram, 84.325, 338
Radiological protection, 5.2–5
    for α-particles, 5.2, 3
    for β-particles, 5.3
    for γ-rays, 5.3
    for neutrons, 5.3
    for X-rays, 5.3
Radio luminography, image plate with, 99.242, 263–265
Radiometric balancing, setting on a line by, 8.30–31
Radio navigation, 85.10–11
    error, 85.15–16
    system, 85.10, 14–15, 38, 53–54, 62
Radon operator, 7.319
Raman effect, 14.122
Raman microspectroscopy, confocal, 14.198–199
Raman scattering, 12.248
Raman spectroscopy, 10.58
Random access memory, sequential digital systems, 89.234
Random number generator, 85.61
Random phase approximation (RPA), 82.205
Random telegraph signal, 87.208, 211
Random vector, 85.18–19
Rank filtering, 3D, 85.179
Rank-order filtering, 89.335; 92.14–18
Rank-selection filter, binary image enhancement, 92.65–68
Rapid freezing technique, Japanese contributions, 96.740–741, 783–784
Rapid kinetics, quantitative particle motion, 98.61–67
Rare earth pole piece electron:
    flux density, 5.226, 227
    lens construction, 5.226
    parameters, 5.227
Raster, 83.204, 225
Rat:
    cornea, 6.194
    cremaster muscle, 6.11, 12, 13, 22

Rat, *continued*
  diaphragm muscle fiber synaptic gutter, 6.205
  hypoglossal nucleus motor neuron, 6.181
  kidney, 6.194
  lachrymal gland, 6.211
  liver, 6.195
  ovary tissue, 6.84
  pancreatic islet, 6.194
  proximal tubules, staining for alkaline phosphatase, 6.185
  spinal chord motor neuron, 6.201, 208
  tissue permeability, 6.217
  ventral horn cell, 6.207
Rate-distortion function, 84.3
Rate-distortion theory, 84.2
Ratemeter:
  advantages, 6.296
  analogue manipulation of signal, 6.297
  for low signal levels, 6.295
  spatial resolution loss, 6.295
  time constant, 6.294
Ratio Imaging, in fluorescence, 14.180–182
Rational realization, 90.111
Rauch–Tung–Striebel algorithm, 85.50
Ray–bounding-box intersection, 85.126–127
Ray intersection effects, electron gun, 8.200
Rayleigh criterion, 10.20; 12.321; 13.268
Rayleigh scattering, 14.122
Rayleigh–Sommerfeld diffraction, 10.139
Rayleigh statistical test:
  domain segmentation, 93.292–293
  orientation analysis, 93.283
Rayleigh waves, 11.162, 163, 164, 181
Rays:
  casting, 85.114, 124, 144
    volume, 85.158
  generation, 85.125
  termination, 85.141
  tracing, 85.124
    charging effects on insulating specimens, 13.77–78
    determination of aberrations by, 13.206–207
    differential equations for, 13.201–203
    direct electron, 13.39–41
    and electron trajectories, 13.173–174, 199–207
    in elliptical defects, 13.56–57
    in magnetic lens, 13.51–54
    numerical methods of, 13.203–207
    paraxial ray equation, 13.205–206
    through Wien filter, 13.110–113
  triode electron gun, 8.139–140
    aberrations, 8.144–147
    asymptotes measurement, 8.147–148
    crossover, 8.140–144, 145–146, 151
    emittance diagrams, 8.152–153, 157–162
    entrance pupil at the cathode, 8.141, 145–147
Raytheon, 91.234, 238–239
RCA:
  commercial microscope development, 96.479
  semiconductor history, 91.149
  Types A, B, 10.222–223
  photography, 10.252
Reactive ion etching (RIE), 83.45
READ/WRITE mechanism,
  single-electron cells, 89.235–237
Real diffuse scattering, 90.348–349
Realizability, 90.117–119
Real process, 86.183
Real-time processes, extrapolation, 94.391–392
Rear port tube, 91.236
Reasoning engine, 86.83, 93, 109–110, 112, 120, 125, 168
Receiver, defined, 99.67
Reception, linear phase sensitive, 84.326
Receptive field (RF):
  Gabor function, 97.41–43
  neuron, 97.40–41
Reciprocal lattice vector, 86.186, 187, 203, 219, 221

Reciprocal rods, 11.79
Reciprocal space, discretization, 7.312
Reciprocity theorem, 11.59
Recombination velocity, charge carriers, 10.60, 64
Reconstructed wave, 12.67, 86
Reconstruction *See* Algebraic reconstruction; Image reconstruction
Recovery time *See* Storage time
Rectangular pixels, 93.229, 272, 274–275
Rectification, maxpolynomials, 90.105–106
Rectifier, history, 91.142
Rectify algorithm, 90.105
Recursive behavior, 94.291–292
  animal and machine behavior, 94.271–272
Recursive dyadic Green's function, microstrip circulators:
  three-dimensional model, 98.219–238
  two-dimensional model, 98.79–81, 98–108, 121–127, 316
Recursive filter, 92.21–22, 57
Recursive formulas, to tenth-order Hamiltonian functions, 91.7, 33–34
Recursive maximum likelihood estimate (RMLE), 84.302–309
Reduced fingerprints, mathematical morphology, 99.32–37
Reduced transform (RT) algorithm, 93.2–3, 16, 31
  affine group, 93.30–31, 39–41
  fast Fourier transform algorithm, 93.16–17, 21–27
  hybrid RT/GT algorithm, 93.25–26
  point group, 93.31–39
  XSUP#sup-invariant algorithm, 93.41–42
Reduct, 94.154
Reductionist approach, brain, 94.262–263, 285
Redundancy, temporal redundancy, 97.235
Redundancy removal, 97.192, 193
Redundant number system, 89.60

Reference particle, 86.236, 241
Refinement, grid, 82.345
Reflectance spectroscopy, photodiode calibration, 99.100
Reflected beam amplitude, 90.229
Reflection:
  fluorescent observation at microscope objective, 14.326–329
  neuroscience, 94.294
  selective, 14.122
  total, fluorescence, 14.178–180
Reflection coefficients, 82.136; 90.318
Reflection electron microscope, 4.163; 10.228–229
Reflection high-energy electron diffraction (RHEED), 11.65; 86.211; 90.210, 241, 243, 317
  equations, 86.176, 205, 210
  from crystal slab, 90.248–250
  from semi-infinite crystal, 90.235, 243–248
  imaging plate system with, 99.286–288
  scattering matrix method, 86.176, 210
  surface resonance scattering, 90.323–334
  tensor RHEED, 90.276–279
  total reflection angle X-ray spectroscopy and, 11.94
Reflection imaging, 11.58
Reflection symmetric residual quantizers, 84039–37; 84.45–50
Reflectivity measurements, 82.201
Reformulation, minimax algebra, 90.86–88
Refractile objects, 6.87, 117
Refractive index, 14.267–269
  gradients:
    controlled variation, 8.6, 8
    determination, 8.11–14
    spatial derivatives, 8.11
  of liquids, 1.69
  match, 1.46, 57
  of specimen, 1.99
Refractometer, 1.57, 59, 70, 75
Refractory materials, 83.17, 18, 54, 57
Refreshed graphics, 91.235, 237

Refresh rate, 86.2, 25
Region, 84.225
Region adjacency graph, 84.225, 229
Region-based analysis, 86.129–130, 1255
Regression, 85.17
Regularity, of partial differential equations, 82.338, 340
Regularization, 97.87–89
  Bayesian approach, 97.87–88, 98–104
  discontinuities, 97.89–91, 108–118
    duality theorem, 97.115–118
    dual theorem, 97.91
    explicit treatment, 97.110–115, 154–166
    implicit treatment, 97.108–110, 166–181
    line continuation constraint, 97.130–141, 142
  edge-preserving algorithms, 97.91–93, 118–129
    extended GNC algorithm, 97.132–136, 171–175
    GEM algorithm, 97.93, 127–129, 153, 162–166
    GNC algorithm, 97.90, 91, 93, 124–127, 153, 168–175
  edge-preserving regularization, 97.93–94
    Markov random fields, 97.90, 105, 106–108
    theory, 97.104–118
  Hopfield energy function and, 87.17
  hyperparameters, 97.141–143
  ill-posed problem and, 87.12 f
  inverse problem, 97.96–98, 99–101
    Gaussian case, 97.103–104
    optimal estimators based on cost functions, 97.101–103
    posterior density, 97.100–101
    prior information, 97.100
    states of information, 97.98–101
  Moore–Penrose generalized inverse and, 87.22
  neural matrix inverse and, 87.26 f
  singular value decomposition and, 87.31

Regularization parameter, 97.96, 143–146
Regular lattices, 97.216
Reichert MeF microscope, 5.20–27
Rejection filters, generating differential equations, 94.381
Relative function, 94.327
Relative information, 91.37
Relativistic potential, 2.169
Relativistic two-particle states
  relativistic singlet state and invariants, 95.361–364
  vectors, 95.361
Relativity:
  general relativity, 90.170–174, 188
  special relativity, 90.147–149, 186
Relaxation microscopy, under fluorescence, 14.196–198
RELAY project, 91.198
REM (software), 86.107, 109–110, 112, 124, 165
Remolded tip, 93.80, 82–83, 106
Remote control microscope:
  accessories, 5.36, 38
  Bausch and Lomb, 5.35, 36
  Brachet "Telemicroscope," 5.36
  choice, 5.38, 39
  commercial model, 5.38, 39
  cost, 5.38, 39
  ease of operation, 5.39
  evolution, 5.20–30
  Leitz MM5 RT, 5.31, 32
  modified standard model, 5.38
  Reichert MeF, 5.20–27
  requirements, 5.20
  Union Optical Company "Farom," 5.32–35
  Vickers, 5.36
  Zeiss Neophot, 5.27–30
Remote sensing image:
  contrast enhancement, 92.28–29
  detail enhancement, 92.45, 46
  extremum sharpening, 92.33, 46
  max/min-median filter, 92.41
  smoothing of SAR speckle noise, 92.13–14

top-hat transformation, 92.64
Replica technique:
  development, 96.150, 281, 338, 440, 462
  Japanese contributions, 96.774–775
  Southern Africa contributions, 96.338
Representation, 84.132; 85.112
  continuous representation, 85.98
  discrete representation, 85.98
  explicit, 84.180
  invariant pattern recognition, 84.131–192
  surfaces:
    Monge patch, 84.168, 189
    parametric, 84.168, 188
  three-dimensional image display, 85.78–220
  uniqueness, 84.132–133
Residual quantizers (RQs), 84.4–6, 11–12, 52
  desing algorithm, 84.37–38
  notation, 84.12
  optimization, 84.12–13
  rate distortion data, 84.52–57
  reflection symmetric residual quantizers, 84039–37; 84.45–50
  vector residual quantizers, 84.26–30
Residuation, minimax algebra, 90.113–114
Residue number system, 89.59
Resin particle composites, 83.32
Resins (embedding), 2.322
Resist, 83.36
Resistor, 87.232, 235, 239
Resolution, 2.47, 108–110
  autoradiography, 2.152, 153; 3.240–242
  biological microscopy, 5.176, 188; 7.286
  by computer synthesis, 4.113–119
  crystal-aperture scanning transmission electron microscopy, 93.91–94
  dark-field electron microscope, 5.302, 342, 343
  defined, 1.130
  digital image processing, 4.130–135
  electron microscope, 2.170; 5.173, 175, 176, 177, 188, 190, 191, 192, 211, 215, 233, 234, 240, 298–303
  field-electron microscope, 2.350
  field-ion microscope, 2.346
  high, problems, 4.86–87
  high-voltage microscope, 2.173, 194
  imaging plate system, 99.257
  improvement over time, history, 96.791–801
  of iron pole piece lens, 5.224, 225
  of iron shrouded solenoid electron lens, 5.220
  of magnetic structures, 5.243
  in microscope objective, 14.254–257
  minimax algebra, 90.99–106
  nature of, 5.242, 243
  orientation analysis, 93.275–276
  phase error and, 7.267
  photographic, 1.198
  photometric, 2.84, 115
  point, 4.67–71
  Rayleigh criterion, 10.20
  reconstruction from finite number of projections and, 7.327, 328
  scanning electron microscopy, 83.206, 215–219, 228
    at low beam voltage, 83.228, 231
    backscattered electrons, 83.238
    improvement over time, 96.525, 530
    limitations, 83.232, 236
    LVSEMs, 83.230–231, 238
    multifactorial approach, 83.241
    optimal, 83.220, 240–242, 253
    simulation, 83.218–219
    specimen position, 83.232
    specimen preparation, 83.232
    theoretical, 83.232, 239
    topographic, 83.219
    ultimate, 83.110
  scanning transmission electron microscopy, 7.199
    improvement over time, 96.524, 526
  Scherzer's limit, 96.797
  of solenoid electron lens, 5.216
  spatial, 85.84, 213
  subatomic, 93.64

Resolution, *continued*
  of superconducting electron lens, 5.202, 211, 234, 235
  two point, conventional systems, 10.21, 22
Resolution algorithm, 90.100–101, 105
Resolution cell, 84.326
Resolution limit:
  of light microscope, 1.116
  of linear lattices, 1.126, 134, 136, 148
  of SAM, 11.174
  for two-image points, 1.121, 143
Resolution parameter, 2.173, 224
Resolving power, 86.234
Resonance detuning parameter, 90.306
Resonance energy transfer, 14.196–198
  using fluorescence, 14.189–191
Resonance scattering, electron diffraction, 90.293–334
Resonant microscope, 10.79–80
Resonant tunneling, 89.126, 128
Resonant tunneling devices, 89.121–142, 222, 245
  applications, 89.135–136
  inelastic scattering, 89.134–135
  reproducibility, 89.222, 244
  space-charge effects, 89.130–134
  spectroscopy, 89.135–136
  transistors, 89.136, 245
  tunneling time, 89.136–142
Resonant tunneling electron spectroscopy, 89.135–136
Resonant tunneling transistors, 89.136, 245
Responsivity, photodetectors, 99.76
Rest mass, general relativity, 90.174
Restoration *See* Image restoration
Restriction function, 90.360
Retardation wavelength, 87.83–86
Retarding grid analyzer, 12.169
Retinal rods, 5.338
Reverse-biased *p-n* junction, 99.172–173, 174, 185, 216–229
RHEED *See* Reflection high-energy electron diffraction
Rhodopsin, bacterial, 8.108
Rhombic stacking, ovals, 7.27

Ribonucleic acid (RNA):
  base damage by electron irradiation, 95.144
  complex formation, 8.127
  dye-binding, 8.53
  Fresnel projection microscopy, 95.132, 134, 136, 138, 144
  secondary structure, 8.127
  single-stranded, 8.127
Ribosomes:
  electron microscopy, 7.288
  negative staining, 7.288
  three-dimensional reconstruction, 7.367
Richardson–Dushman equation, total emission current density, for, 8.167, 178
Richardson's thermionic emission equation, 8.212, 213; 83.26
Richter's condition, 6.57, 96, 99
Richtstrahlwert, 8.139
Ripple, high voltage supply, 1.125
Ripple structure:
  in thin permalloy film, 5.290, 291
  in thin polycrystalline films, 5.240, 289, 290, 291
Risk for estimation, 97.144
Risk for protection, 97.144
RMLE *See* Recursive maximum likelihood estimate
RNA *See* Ribonucleic acid
RNA polymerase, 8.126
Roberts operator, 89.87–88, 90, 91; 93.235, 236, 266, 277, 323
Robinson detector, development, 96.51
Robustness, minimax algebra, 90.76
Rocking curves:
  CBED, 90.211, 218–221
  dynamical theory of diffraction, 93.60
Rock textures, optical transforms, 7.34
Ronchi-grid, 8.12, 13
Roof edge, 93.231
Rosenbrock's banana, 83.150
Rosen–Margenau–Page potential, 98.11
Rosette diagram, 93.240, 241–243, 245, 254–255, 279

Rosin–Rammler–Sperling distribution equation, 5.152
Rotating glass block:
   apparatus, for flow velocity, 1.11, 31, 36
   eyepiece apparatus, for high flow velocities, 1.17, 33
Rotating-prism scanner, 8.31–32
Rotational symmetry, two-dimensional images, 7.302
Rotation matrix, 85.12, 29
RoughClass system, 94.184
RoughDAS system, 94.184
Rough definability, 94.166–171
Rough measure, 94.182, 189
Rough set, defined, 94.165–166
Rough set theory, 94.151–194
   applications, 94.182–194
      attribute significance, 94.182–183
      LERS, 94.184–193
      real-world applications, 94.193–194
   concepts and definitions:
      certainty and possibility, 94.166
      indiscernibility, 94.152–157
      lower and upper approximations of a set, 94.157–165
      rough definability, 94.166–171
      rough set, 94.165–166
   and evidence theory, 94.171–182
      basic properties, 94.171–176
      Dempster rule of combination, 94.176–180
      numerical measure of rough sets, 94.181–182
Round compound field electron emission lens, 8.243–244
Rounding error, 85.32
Roundness, 3.36, 64
Row–column algorithm, 93.45
Rowlinson potential, 98.45
RPA *See* Random phase approximation
RQs *See* Residual quantizers
RRS function, 5.152
RT algorithm *See* Reduced transform (RT) algorithm
Rule induction, LERS, 94.184–194

Runge–Kutta–Fehlberg formula, 13.202
Runge–Kutta formula, 13.41
   and ray tracing, 13.202–203, 205
Runge–Kutta–Verner method, 85.60
Ruska, Ernst:
   career, 96.132–137, 416–418, 639, 792–793
   children, 95.26, 59–61
   death, 95.61
   electron microscope and, 96.11–12, 65, 132–135, 416, 462–463
   Ernst Ruska Prize, 96.158
   extramural activities, 95.47–48
   family background, 95.4–13
   Knoll's influence, 95.13–14, 16–18, 39, 42
   magnetic lens development, 96.134
   marriage, 95.22–23, 59–61
   Max Planck–Gesellschaft experience, 95.37–38
   military service, 95.23
   minimization of external vibration and, 95.44–47
   Nobel Prize, 95.53–54, 56–58; 96.131, 145, 159, 794
   politics, 95.48–50
   postwar experiences, 95.28–34
   retirement, 95.50–52
   scanning electron microscope, 96.635–638
   Siemens experience, 95.25–28, 30, 35–37
   single-field condenser-objective development, 95.41–42
   Soviet relationship, 95.28–30
   Technische Hochschule Berlin experience, 95.13–14, 16–18
   transmission electron microscope development, 95.4, 18, 20, 25, 37
Russia:
   commercial microscope development at Sumy, 96.428–429, 437–438, 454, 460, 469, 471, 482–484, 492, 494–496, 499–500, 502–503, 510, 516–517, 532

Russia, *continued*
  electron diffraction camera production, 96.488, 500–501
  electron microprobe production, 96.465
  pioneers, 96.523, 574

S

Saccharomyces cerevisiae, 5.333
SAED *See* Selected area electron diffraction
SAGCM avalanche photodiodes, 99.73, 92–94
  critical device parameters extraction, 99.102–120, 153, 155
    error analysis, 99.110–113, 118–119, 155
  photogain, 99.74, 120–135, 155
    temperature dependence, 99.135–150
  planar, 99.96–102, 121
SAGM InP/InGaAs avalanche photodiodes, 99.90–92
Sakaki, 12.22
Salts, structural conservation, use in, 8.116–117
SAM, 11.153, 175
SAM InP/InGaAs avalanche photodiodes, 99.90–92
Sampling:
  analog-to-digital converting, 84.2
  vision modeling, 97.3–4, 45–50
  visual cortex, 97.45–50
  volume, 84.322
Sampling interval, electron off-axis holography, 89.23, 24
Sandstone, transforms, 7.63
Sanidine phenocrysts, 7.37
Sapphire lens, 11.154
SAR *See* Synthetic aperture radar
Satellite communications, 91.198–199
Saturated-absorption spectroscopy, 8.44
Saturation density, photographic, 1.184
SA-W-LVQ *See* Successive approximation wavelet lattice vector quantization

SBTF *See* Sideband transfer function
Scalar field, 85.89, 99
Scalar information, 90.135
Scalar measurement, 85.21–24, 36
Scalar potential:
  electric, 82.11, 26, 41
  magnetic, 82.27, 42
  reduced, 82.12
  total, 82.13
  modified, 82.54, 61
Scalar power series, 90.80–83
Scalar quantization, 97.193, 194
  finite state scalar quantization, 97.203
  wavelets, 97.200–201, 202–203
Scalar quantizers, 84.6
  residual, 84.14–37
  single-stage, 84.6–9
  trellis-coded, 84.51
Scalar residual quantizers, 84.14–26
Scalars, 90.5, 10, 23–24
Scalar theory, charged-particle wave optics, 97.316–317
  axially symmetric electrostatic lenses, 97.320–321
  axially symmetric magnetic lenses, 97.282–316
  electrostatic quadrupole lenses, 97.321–322
  free propagation, 97.279–282
  general formalism, 97.259–279
  magnetic quadrupole lenses, 97.317–320
Scale-space:
  defined, 99.2
  Gaussian scale-space, 99.3–5, 55
  gradient watershed region, 99.51–53
  morphology, 99.55
    fingerprints, 99.4–5, 29–37, 55
    future work, 99.56–57
    limitations, 99.55–56
    multiscale closing–opening scale-space, 99.22–29
    multiscale dilation–erosion scale-space, 99.16–22
    multiscale morphology, 99.8–15

for regions, 99.46–53
signal extrema, 99.20–22
structuring functions, 99.37–46
Scale-space filtering, 99.2
Scaling, 85.118; 86.27
Scandinavia:
  electron microscopy:
    cell ultrastructure, 96.307–308
    enzyme cytochemistry, 96.308–309
    growth, 96.306–312, 314
    materials science, 96.310–311
    methacrylate embedding method, 96.306
    ultramicrotome, 96.306, 310
  Scandinavian Society for Electron Microscopy, 96.315–320
  Siegbahn electron microscope:
    characteristics, 96.303–304
    commercial production, 96.304–306
    development, 96.301–302
Scan Line Array Processor, 87.292
Scanners:
  beam-deflecting, 6.136
  electromagnetically deflected mirror, 6.164–168
  flying spot, 6.137
  glass cube, 6.138
  mirror drum, 6.137
  prismatic, 6.137–142
  refracting, 6.137, 138, 139, 140, 141
  television camera, 6.30, 42, 137
Scanning, 84.329
  front-to-back, 85.156
  in image analysis, 4.362–363
  incremental, 85.120
  large drawing, 84.254
  measurements, 1.98
  microscopy and, 1.101; 83.203–204
    mechanical, 83.204
    scanning speed, 83.234–235
    types, 83.206–207
  photoelectric setting microscopes, 8.35–37
  recursive, back-to-front, 85.122
  slice-by-slice:
    back-to-front, 85.121, 143
    front-to-back, 85.123, 143
Scanning acoustic microscopy, 10.70
Scanning electron microscopy (SEM), 2.42, 43, 62, 74, 80, 142; 4.162, 244; 5.41; 6.277; 83.113, 203–273
  advantages, 83.236–237
  backscattered electrons, 83.205
  beam voltage, 83.207, 209, 220, 241
  contrast, 83.209, 235–236
    effects beam misalignment, 83.230
    radiation damage, 83.209–210
    resolution, 83.220, 241
  cathodoluminescent, 83.213
  CD measurement in, 13.127
  charging, 83.233–235
    at low beam voltage, 83.234
    avoiding, 83.233–234, 254
    insulating specimens, 13.74–92
  collection system, 13.113–119
  colloidal gold labeling, 83.214–216, 226–227
  contamination deposited, 10.63
  contrast, 83.205–206, 208–209, 213, 235
  cryo-SEM, 83.252–253
  deflection elements, 13.113–119, 179–180
  deflection fields, 13.146–152
  density and, 83.206, 209, 215–216, 218, 225, 227, 231, 233, 247–248, 249, 251
  depth of information, 10.292–294
  development, 13.124–125; 96.142–143, 466–468, 638–639, 643–644
    European, 96.466–470, 501–506, 509–513
    Japanese, 96.704–705, 712–713
  direct and photographic, 2.83, 84
  EBIC method, 10.69
  electron interaction in, 83.205–206, 217–219, 234
  electrostatic charging and, 13.231
  environmental, 83.21
  field, 83.122
  focusing system, 13.113–119

Scanning electron microscopy
(SEM), *continued*
   frozen-hydrated SEM, 83.252–256
   future developments, 83.255, 259
   high resolution, 10.249
   high voltage, 83.209, 236
   history, 10.226–227; 83.204–205, 256–257, 258
   image processing and restoration, 13.295–298; 83.237–238
   inspection based on, 13.127
   lenses, 12.101–102
   limitations, 83.206, 220
   living specimens, 83.209
   low-temperature SEM, 83.213–214, 252–253, 252–256
      contamination, 83.213
      cryo-preparation, 83.252–255
      freeze-fracture/cryo-SEM comparison, 83.237–238
      freeze-fracture/thaw-fix, 83.242, 245
      frozen-hydrated SEM, 83.252–256
   low-voltage SEM *See* Low-voltage scanning electron microscopy
   magnetic immersion lenses and, 13.92–107
   magnetic microstructure, 98.331–332
   magnification, 83.204
   method, 13.76–78
   optimum conditions, 83.236–237, 241, 247, 253
   origins, 10.220
   photoelectron microscopy, 10.291–293
   probe size, 13.292–295
   recording, 10.253
   resolution, 83.206, 215–219, 228; 96.525, 530
   scanning speed, 83.234–235
   secondary electrons, 83.215–219
   semiconductor *See* Semiconductors
   signal types, 83.205
   solid state detector, 6.281
   sources *See* Electron sources
   stereo SEM images, 83.245–246, 248, 250
   surface photovoltage, 10.322
   through the lens detection, 12.103–106
   topographic image, 83.210, 218, 220, 235
   transmitted, 83.204, 247
   vacuum, 83.210, 228, 259
   and Wien filters, 13.107
Scanning electron microscopy with polarization analysis (SEMPA), magnetic microstructure, 98.333
Scanning interferometers, 8.26, 33
Scanning microscope photometer, 5.125, 137
Scanning mirror electron microscopy, 4.227–232, 249
Scanning near-field optical microscopy (SNOM), 12.243–244
   applications, 12.246–252
   experimental basis, 12.272–309
   history, 12.247–248
   principles, 12.244–246
   theory, 12.252–272
Scanning optical microscopy, 9.327–329, 334; 10.20, 22
   acoustic microscope, 10.70; *See also* Confocal imaging
   advantages, 10.8–14, 69–70
   antireflection coating, 10.65
   beam scanning, 10.70–81
      noise and flare, 10.74
   confocal microscopy, optical system, 10.2, 7
   diagram, 10.3
   differential phase contrast, 10.49–55
   diffusion length measurement, 10.60–61
   flying spot system, 10.2, 61–62
      advantages, 10.70
   heterodyne microscopy, 10.82–84
   image analysis, 10.9
   image processing, 10.9–14
   interference microscopy, 10.75–81
   laser source microscopy, 10.81–87
      CRT, 10.70
      heterodyne, 10.82–84
      nonlinear, 10.84–87
   mechanical scanning, 10.62

Nipkow wheel, 10.16, 61
noise and flare, 10.2, 73–75, 74
nonlinear microscopy, 10.84–87
OBIC methods, 10.61–65, 68–69
   EBIC/OBIC dislocations, 10.62–66
object-scanning, 10.70–81
   advantages, 10.72
one-dimensional, chopped beam, 10.59–62
optical generation of charge-carriers, 10.59–70
optical systems, 10.2
Oxford model, 10.3–4
penetration of electrons, 10.64
point detector scanning, comparison, point-source, 10.49–70
recombination velocity, 10.60
reflected light, and OBIC images, 10.64–66
rotating mirrors, 10.70
second-generation harmonics, 10.84–89
single moving mirror, 10.62
spectroscopic microscopy, 10.55–58
stage design, 10.73
summary and conclusions, 10.87–90
two-dimensional scanning, 10.61–64
Scanning photoelectric microscope, 6.135
Scanning reflection electron microscopy, 11.70
Scanning susceptibility microscopy, 87.183–188
   application, 87.184
   complete flux expulsion model, 87.186–187
   force and compliance-versus-distance curves, 87.185–186
   probe-induced vortex nucleation process, 87.186–187
   relative force variations, 87.188
   sensor, 87.183–184
   total repulsive force, 87.184
Scanning tip lithography (STL), 89.95, 97
Scanning transmission electron microscopy (STEM), 7.101; 8.110; 11.45; 12.96; 83.113; 90.289; 93.57; 94.197
   aberrations, 93.86–87
   crystal-aperture STEM, 93.57–107
   direct imaging of nucleus, 93.87–89, 104
   experimental, 93.66–87, 90–91
   imaging, 93.58–59, 63–66, 87–90, 94–106
   resolution, 93.91–94
   theory, 93.59–66
   currently built instruments, 10.256
   development, European, 96.506–508, 519–520, 644–645
   electron microscopy and, 7.147
   heavy atom discrimination in, 7.192
   holography, 98.373–387, 422–423
   illumination, conical, 7.147–148
   image formation, 94.221–231
   magnetic microstructure, 98.335–337, 342, 353, 357, 358, 360
   microfabric analysis, 93.222, 231, 276, 303, 304
   off-axis holography, 94.232–244, 232–253, 252–253
      Fourier transform, 94.244–246
      spectral signal-to-noise ratio, 94.246–252
   recording, 10.253
   support film contamination and, 128
   Type 1, geometry, 10.19–20
Scanning transmission electron microscopy holography, 98.373–387, 422–423
Scanning tunneling microscopy (STM), 11.74; 12.244; 83.87, 88; 87.49; 89.236, 237
   invention, 96.799
   Japanese contributions, 96.719, 721
Scanning X-ray absorption microscopy, 10.322
Scanning X-ray photoelectron microscopy, 10.321–323
Scattered wave, 90.208
Scattering *See* Electron scattering
Scattering amplitude, 90.207–209
Scattering cross-section, 90.207–209

Scattering matrix, 86.181, 183, 211, 215, 216; 90.228–229
Scattering parameters, three-port circulator, 98.117–120, 238–245, 302
Scattering theory:
  Born approximation, 95.313
  Coulomb scattering, 95.313–314
  Mott scattering, 95.314–315
  spacetime algebra, 95.312–315
Scherzer defocus, 11.32
Scherzer focus, 12.49
Schist, transforms, 7.45
Schlieren techniques, 8.6, 11–14
Schottky barrier, height, 83.22, 24, 71
Schottky-barrier photodiode, 83.31; 87.226, 232; 99.78–79
Schottky devices, 10.68
  effect, photoelectron microscopy, 10.381
Schottky effect, 8.210
Schottky emission cathode, 95.68–69
Schottky's theorem, 12.173
Schrödinger, Ernst:
  cat experiment, 94.263–264
  Vedic cognitive science, 94.264–265, 267
Schrödinger's equation, 13.248–251; 89.112; 90.124; 94.294–295
  in classical limit, 5.270
  electron diffraction, 90.215
  electron optics, 93.174, 178, 184, 185, 187, 190, 200–202, 206, 208–209, 211
  nonrelativistic, 5.269
  paraxial, 93.194, 206
  physical information, 90.161; 94.164
  semi-classical approximation, 5.246–249
  solution:
    and current density, 13.254–255
    electron charge, conservation of, 13.254–255
    partial derivatives and Jacobian determinant of, 13.251–254
Schur complement, 85.39–42, 50, 52
Schur-convexity, 91.48, 67–68, 70

Schwarz–Hora effect, 7.169, 171
Scintillator for electron detectors, 83.223–227
  BSE detector, 83.223–224, 226
Screening length, 82.204, 225, 230, 237
Screen microkymography, 7.76, 79
  quantitative evaluation, 7.89
  registration principles and techniques, 7.78
Screen photomicrography, 7.76
Screen photomicrokymography, registration principles and techniques, 7.78
Screw dislocation, 11.88
Scripts, neuroscience, 94.290
S-curves, 12.221, 231
S(D) operators, 94.335
Seasoning, vacuum microelectronics, 83.6
Sea urchin egg, image, 83.253–256
Secondary electrons, 83.205–206, 208, 216–219
  coefficient, 83.208, 234
  collection field, 83.223–224, 233
  computation, 13.82, 89
  deflection fields and, 13.146
  detector:
    Everhart–Thornley, 83.223
    TEM/SEM, 83.222
  electrostatic charging and, 13.225–226
  emission, 12.155
  high-beam energy scan, 13.82
  imaging, 83.216
  low-beam energy scan, 13.89
  in low-voltage inspection, 13.127
  Monte Carlo, electron scattering simulations, 83.218
  performance, 83.230
  production, 83.216–219
  trajectories, 13.204
Secondary ion mass spectrometry (SIMS), 10.303–304, 305; 11.101, 101–103, 102
  applications, 11.147–148
  high-resolution SIMS, 11.121
  ion sources, 11.103–104

field ion sources, 11.110–112
plasma sources, 11.104–106
surface ionization sources, 11.106–109
mass separation of primary ions, 11.113–120
photodiode calibration, 99.100–101
secondary ion collection, 11.140–147
secondary SIMS with high lateral resolution, 11.120–121
    ion emission microscopy, 11.121–129
    ion microprobes, 11.129–140
Secondary ions, collection, 11.140
Secondary scattering, 88.151
Secondary spectrum, correction of chromatic aberration, 14.309–320
    axial, 14.276–277
Second fundamental form, 84.173, 190
Second law of thermodynamics, information approach, 90.189
Second-order Hamiltonian function, power-series expansion, 91.8
Second-order perturbation, 90.254–255
Second zone lenses, 10.236–237
Section, of bounding surface, 82.5, 6, 7, 8, 9, 10, 11, 16, 17, 19, 26, 28
Section thickness, 3.228–238
Sector magnet, 11.114
SEC tubes, 9.32, 40, 41, 44
Sediment, microfabric, orientation analysis, 93.221–224, 231
Seeing, 85.81
SEGEN (software), 13.76
Segmentation, 84.224
Seidel aberrations, 14.257–270
Seismic studies, 85.80
Selected area electron diffraction (SAED), 11.44; 90.214
Selective adsorption, macromolecules, 8.114–115
Selective extraction, 2.259, 325, 331
Selective reflection, 14.122
Selenium, 91.142
Selenium cells, 4.395, 397
Self, philosophy, 94.294

Self-adjointness, 92.134
Self-coherence, 7.108
Self-distance, 90.144
Self-duality, mathematical morphology, 89.354–358
Self-energy, of holes, 82.210, 212, 217
Self-scanned diode arrays, 8.31
Selvage, crystal, 90.241
Selvage scattering, 90.241, 245–248
SEM *See* Scanning electron microscopy
Semantic label, 84.230
Semiconductor laser, 91.179
Semiconductor photodetectors:
    avalanche photodiodes *See* Avalanche phototiodes
    metal–semiconductor–metal photodetectors, 99.79
    photoconductive detector, 99.79–80
    phototransistors, 99.80
    p-i-n photodiode, 99.77–78
    p-n photodiode, 99.77
    Schottky-barrier photodiodes, 99.78–79
Semiconductor quantum devices, 12.148, 162; 89.93–245
    Aharonov–Bohm effect-based devices, 89.106, 142–178
    connecting on a chip, 89.208–243
    electron wave devices, 89.99–120
    directional couplers, 89.193–199
    granular electronic devices, 89.203–208, 245
    quantum-coupled devices, 89.208–243
    reproducibility, 89.222
    resonant tunneling devices, 89.121–142, 222, 245
    scanning tip lithography, 89.95, 97
    spin precession devices, 89.106, 199–203
    transistors, 89.106, 157, 160–165, 167, 189, 244
    T-structure transistors, 89.178–193
Semiconductors:
    band structure effects, 8.218
    BSE detectors, 83.225
    charging, 83.234

Semiconductors, *continued*
  compensated, 82.206, 230, 233, 244
  doped, 83.7, 20, 24, 71
  doped compensated, 82.207
  Group III–V compounds, 91.171–188
  heavily doped, 82.199
  history, 91.142–166, 146, 149–151, 166–169, 171–188
    United Kingdom, 91.149–151, 168–169, 177–178, 181
    United States, 91.168–169
  infrared technique, 10.8
  integrated circuits *See* Integrated circuits
  lightly doped, 82.198
  low voltage SEM, 83.227–228
  LVSEM, applications, 83.227–228, 249–250
  magnetoplasma studies, 82.86; 92.81–83, 89, 116
  mesoscopic devices, 91.213–227
  mirror electron microscopy, 94.96
  moderately doped, 82.200
  modulation-doped semiconductor heterojunction, 91.214
  optical generation of charge carriers, 10.59–70
  photoelectron microscopy, 10.300–305
  product group, 94.20
  quantum theory, 91.142
  radiation damage, 83.228
  tensor products, 94.20–21
  undoped, 83.7
  use of SEMs in, 13.125, 127, 153
  vacuum microelectronics, 83.7–10, 19, 24, 31, 67, 68, 73
Semidirect product group algebra, 94.19–22
Semi-group properties, structuring functions, 99.15, 37–38
Semi-infinite crystals, 90.235, 243–248
Semi-lattice ordered group, 84.66
Semi-lattice ordered semi-group, 84.68
SEMPER (software), 93.229, 319, 322
Sénarmont compensator, 6.97, 99, 111
Sénarmont OPD measurement, 6.114

Sensitivity, imaging plate system, 99.254–257, 262
Sensors, 83.86
Sensor transmission high-energy electron diffraction (THEED), 90.255–256
Sensory perception, 85.80
Sequential resonant tunneling, 89.129
Serial mass spectrometer, 89.430
Serial photography, movement determination by, 7.73
Serial transformations, mathematical morphology, 89.377–381
Series expansion, 86.255, 264
Servo-controlled focusing system, 6.28, 30, 38–45, 46
  drive system, 6.43, 44
  focus sensing, 6.39–42
  signal processing, 6.42
"Servo force," 87.129
Servo system:
  drive, 6.43, 44
  microscope system, 6.40, 41
  signal processing, 6.42
Sessile drop, 98.42–44
Set mapping; *See also* τ-mapping
  filtering properties, 89.349–358
  mathematical morphology, 89.327–328
  translation-invariant, 89.358–366
Sets, 85.148, 205, 233
  definability, 94.166–171
  definable, 94.158–159
  identity transformation, 85.234
  lower and upper approximations, 94.157–165
Set theory:
  fuzzy set theory, 89.255–264
  image recovery problem, 95.159–160
Setting on a line:
  fixed image scanning by, 8.30–35
  harmonic detection by, 8.33–34
  radiometric balancing by, 8.30–31
  time-measuring techniques by, 8.32–33
Shading, 85.114, 217
  depth, 85.133, 198
  gradient, 85.135, 137

normal-based, 85.134, 135
Z-buffer, 85.134
Shading compensation, 92.20–26
　background extraction, 92.22–24
　rank-order statistics, 92.25–26
　weighted unsharp masking, 92.24–25
Shading-off effect, 6.58, 59–61, 87, 88, 90, 98, 101, 104, 129
Shadow-casting, optical symbolic substitution, 89.84–87
Shadow curves:
　breadth, 8.151–152
　electron gun rays analysis, 8.150
　emittance diagram conversion, 8.152–153
　experimental determination, 8.153–157
　shape, 8.148–150, 153–155
Shadow-electron microscope, 10.229
Shadowing, 85.114
Shadow mask color tube, 91.233, 235–236
Shadow projection mirror EM, 4.249–259
Shannon entropy, 90.125
Shannon–Gibbs inequalities, 91.38, 42, 53–57
Shannon's entropy, 91.37, 38, 96
Shape coefficients, 3.35
Shape estimation, 88.251
Shape operator, 84.191
Shape representation, algebraic, 86.149–151, 164
Shapiro matrix, 89.181, 183
Shearing interference system:
　construction, 6.107, 110
　differential interference, 6.110
　polarizer, 6.112
Shielded cells, 5.5–8
Shift-invariant case, 90.135–136
Shimadzu Corporation:
　early research interests, 96.668–670
　SM-1, 96.665–666, 670
　SM-1A, 96.666–667
　SM-1B, 96.667
Shockley–Read–Hall generation, 86.10
Shortest path problem, 84.64
Short-focal-length lens, 83.123

Short-term memory, 94.289–290
Short-time Fourier transform (STFT), signal description, 94.320, 321
Shrinking, binary image components, 90.389–396, 407–408
Shrinking spiral path, 90.375
Sickle cell anemia, French research, 96.95–96
Sideband holography, 7.167, 168
Sidebands, 12.35
Sideband transfer function (SBTF), 94.219–220, 246–252
Sidewinder missile, 91.174
Siemens Company:
　electrostatic astigmatism corrector, 96.446
　Elmiskop 1, 96.454–455, 462–463
　Elmiskop 2, 96.443
　Elmiskop 51, 96.482
　Elmiskop 101, 96.480
　Elmiskop 102, 96.495
　Elmiskop CT, 96.150, 495–496, 529
　prototype electron microscope development, 96.418–422
　semiconductor history, 91.172–173, 176, 177
　ST 100F, 96.507, 529, 791
　Übermikroskop, 10.222, 223, 252; 96.423–429, 436
　ÜM 100, 96.436, 446
　withdrawal from microscope production, 96.528–529, 791
　World War II impact, 96.427–428
SIGMA (software), 86.116, 160
Sigma band, 85.55
Signal:
　defined, 94.316, 317
　density resolution, 99.261–262
　description, wavelet transforms, 94.321
　detection:
　　statistical theory, 84.338
　　structural properties method, 94.378–379
　fingerprint, 99.4–5

Signal, *continued*
  identification, structural properties method, 94.383–386
  modeling, 88.311
  redundancy, 97.192
  theory, 94.315–317
Signal description:
  criteria for comparing methods, 94.325–327
  defined, 94.317
  direct description, 94.318
  equivalent linear time-invariant discrete systems, 94.322
  expansion in a series of orthogonal functions, 94.319
  generalized phase planes, 94.324–325, 347–365
  higher order spectra, 94.322–324
  integrated transforms, 94.320
  mixed time–frequency transforms, 94.320–321
  phase and envelope, 94.318
  structural properties method, 94.327–347
    applications, 94.365–394
    generalized phase planes, 94.324–325
    generating differential equations, 94.324, 327, 336–347
    state function, 94.324, 327, 329–335
Signal processing, 97.192
  compression, 97.51, 192–194
  digital coding, 97.192–194
  feedforward networks, 94.277
  Gabor functions, 97.5
  group algebra, 94.46
  nervous system, 94.274–275, 279
  phase trajectories, 94.353–365
  separation, 94.376–399
  theory, 97.2–3
Signal-to-interference ratio (SIR), 84.265–271
Signal-to-noise ratio (SNR), 1.111; 12.175, 227; 82.100; 83.208–209; 84.329, 333
  analog optical fiber communication, 99.68
  scanning transmission electron microscopy, 7.147
  segmental, 82.101
  spectral SNR:
    bright field imaging, 94.220
    STEM off-axis holography, 94.246–252, 253
    TEM off-axis holography, 94.218–221
    transmission electron microscopy, 94.211–213
Signal-to-quantization noise ratio (SQNR), 84.38
Signal uncertainty, 97.10
Signed digit arithmetic, optical symbolic substitution, 89.59–71
Silicon, 83.20, 21, 33, 43, 46
  amorphous films, lattice fringes, 7.156
  avalanche photodiodes, 99.89
  cones, 83.45
  emitters, 83.24
  heavily doped, n type, 82.252
  Johnson limit, 89.243
  lightly doped, 82.206
  moderately doped, 82.225
  properties, 92.82
  semiconductor history, 91.147, 150, 171, 173, 176, 178
  substrate, 83.29
Silicon carbide, 86.72
Silicon detector, Li-drifted, 6.281
Silicon diodes, 9.281–282
Silver halides, dissociation, 8.17
SIMD architecture *See* Single instruction stream multiple data stream architecture
Similarity relations, 89.294–296
Simple linear dependence, minimax algebra, 90.36
SIMS *See* Secondary ion mass spectrometry
Simulated annealing, 87.4, 87.16, 87.25 et seq.
Simulated annealing minimization

algorithm, 97.120–122, 155
Simulated flight data, 85.57–61
Simulated line scan:
   charging effects on insulating specimens, 13.78
   high-beam energy scan, 13.85–86
   low-beam energy scan, 13.86, 91–92
Simulations *See* Computer simulations; Numerical simulations
Simultaneous iterative reconstruction techniques (SIRT), 7.327; 95.177, 183
Single atoms, direct images of, 8.238
Single-crystal cathode tips, orientation, electron beam brightness measurement and, 8.183–186
Single crystals, 83.12, 13, 19, 46
   optical diffraction analysis, 7.33
   semiconductor history, 91.143
   tips, 83.33
Single-electron excitation, 86.187
Single-electron inelastic scattering, 86.186
Single-electron logic devices, quantum-coupled spin-polarized, 89.223–235, 241–243
Single-field condenser-objective, 1.137
   advantages of, 1.146
   cross-section of, 1.146
   development, 95.41–42
   field distribution, 1.143
   lattice resolution of, 1.148
   parameters of, 1.145
   permissible disturbances in, 1.147
   point resolution of, 1.144
   ray paths in, 1.139
Single Global Covering algorithm, 94.188–190, 191
Single-hit law (photographic), 1.189, 191, 193
Single-impact ionization, 86.186
Single instruction stream multiple data stream (SIMD) architecture, 87.262, 272, 274, 277–279, 287, 294–295

SIMD mesh-connected computers, parallel image processing, 90.353–426
Single Local Covering algorithm, 94.190–191
Single-point image:
   aberration-free lenses, 10.21
   formula, 10.20
   intensity, 10.17
Single pole lenses, 10.16–19, 237–238; 12.101
Single-shell lattice codebooks, 97.218
Single sideband holography, 7.220, 251, 252, 254, 255
Single slot, surface field, 92.108–109
Single-stage quantizers, 84.6–11
Singularity:
   of currents, edge (end), function of, 82.2–3, 13–15, 17–18, 41
   feed (attachment), 82.2–3, 13–15, 17–18, 41
Singular location, 84.248
Singular matrix, 85.33–34
Singular value analysis, 87.13
Singular value decomposition (SVD), 85.32–34, 85.36, 85.62, 87.22, 87.27, 87.31 eq seq., 87.34
Sinusoidal transforms, 88.1, 10–15
   integer cosine transforms, 88.51–68
   integer sinusoidal transforms, 88.24–41
   orthogonal transform, 88.5–10
   Walsh transforms, 88.15–24
SiO replicates, 8.85–93
   analysis and interpretation, 8.97–99
   feasibility studies, 8.77–79
   low vapor pressure liquids, of, 8.79
   microsurface spreading and, 8.93–97
   nucleation, wet surfaces on, 8.79
   water drops, 8.77–79, 82
   wet ferritin particles, 8.85
   wet specimens, 8.97–103
SIP microscopes, 8.32–33
   discrimination, 8.38
   line position measurement, 8.39
   nonlinearity of output, 8.40–41

SIP microscopes, *continued*
  scanning mechanisms, 8.35
SIR *See* Signal-to-interference ratio
SIRT *See* Simultaneous iterative
  reconstruction technique
Sixth-order Hamiltonian function,
  power-series expansion, 91.8–9, 13
Size analysis, optical microscopy:
  arithmetic of, 3.82–85
  errors, 3.88–90
  limits of, 3.42
  in sections, 3.85–88
  semi-automatic, 3.49
  techniques, 3.42
Skeletonizing technique, 84.115
  data compression and, 84.120
  image algebra notation, 84.119
  matrix notation, 84.115
Sketchpad system, 91.235
Slab crystals, 90.229–230
  RHEED, 90.248–250
Slab stabilization, drop formation,
  98.10–12
Slater–Kirkwood potential, 98.12
Sliding Memory Plane, 87.289
SLOR *See* Successive line over-relaxation
  methods
Slot, image processing, 86.117–118, 154
SLS *See* Strained layer superlattice
Small-angle approximation, 86.194, 195
Small-angle electron diffraction *See*
  Electron diffraction
Smallest open neighborhood (SON),
  84.205
Small molecules:
  early structure analyses, 88.115
  lattice images, 88.158
  structure analyses, examples of, 88.157ff
  structures by direct methods, 88.139
Small parameter, 86.252, 263
Smalltalk, 88.68, 77
Smearing:
  in microtransforms, 7.48
  microtransforms, 7.41
  in optical diffraction analysis, 7.33

Smell, 85.81
Smetic liquid crystals, 91.247
Smoothing, 88.303, 304; 92.9–19
  with additive noise or texture, 92.10–13
  filter, 88.322, 323
  with multiplicative noise or texture,
    92.13–14
  rank-order filtering, 92.14–18
  adaptive quartile filter, 92.14–16
  center-weighted median filter, 92.17
  composite enhancement filter,
    92.16–17
  iterative noise peak elimination filter,
    92.17–18
  streak suppression, 92.19
Smoothing algorithm, 85.49–50
Smoothing filter, 84.342–343
Smoothing property, 82.336
Smoothness, image processing, 97.89, 97
Snell's law, 14.269; 95.317
SNOM *See* Scanning near-field optical
  microscopy
"Snorkel" lens, 10.238
SNR *See* Signal-to-noise ratio
Sobel edge detector, 88.304
Sobel operator, 93.235, 236, 253, 261–262,
  266, 267, 269, 277, 323
SOC *See* Sphere-on-orthogonal-cone field
  emission model
Social computing, 94.263
Sodium chloride, mirror electron
  microscopy image, 4.242–243
Sodium silicate support films, 8.123
Soft algebra, fuzzy set theory, 89.256
Soft erosion, 92.49
Soft error, 86.28
Soft iron, 5.147, 148, 158, 159
Soft morphological filter, 92.49–52
Software; *See also* individual software
  programs
  accuracy of, 13.48
  bi-potential lens, spherical aberration of,
    13.48–51
  for direct electron ray tracing, 13.39–41
  for FDM, 13.31–32

for spherical capacitor tests, 13.45–48
testing, 13.44–59
for three-dimensional systems, 13.25–44
tolerance calculations, 13.54–59
SOIL *See* Swing objective immersion lens
Soils:
  microfabric, orientation analysis, 93.221–224, 226–227, 259, 275–276, 281, 318–319
  use of image analyzing computer, 4.380
SOL *See* Swing objective lens
Solenoid, thin in magnetic lens, 13.140–141
Solenoid electron lenses, 14.18–19
  aberration, 5.216
  disadvantages, 5.217
  electron optical power, 5.217
  flux density, 5.215, 216
  resolution, 5.216
  sequence of coils, 5.217
Solid–liquid interfaces
  macromolecules interaction with, 8.113–114
  properties, 8.111–113
Solid-state physics, use of mirror EM, 4.247–248
Solid–vacuum transition, 87.50–51
Solution method, image recovery problem, 95.157–158
Solvation forces, 87.112–119
  Clausius–Mosotti equation, 87.115
  dielectric permittivity, 87.115
  excess near-surface molecular density, 87.112–113
  force per unity probe radius, 87.117–118
  oscillating attractive/repulsive interaction, 87.118
  oscillating Hamaker constant, 87.116
  oscillatory, 87.114
  periodic molecular ordering, 87.116–117
  probe-sample interaction, 87.114–115
Somatosensory cortex, experiments, 94.291
Sommerfeld's free electron gas model, 8.216

Sommerfeld theory, 12.242, 254
SON *See* Smallest open neighborhood
Sonoluminescence, 14.122
Soot layers:
  absorption coefficient, 6.68
  for negative phase ring, 6.70
  optical path difference, 6.67
  optical properties, 6.66–68, 71
  optical thickness variation, 8.20
  preparation, 6.66
  refractive index, 6.67, 68, 70, 78
Soot phase ring:
  amplitude ring, 6.82
  curting, 6.68, 69
  development, 6.65, 66
  durability, 6.68
  from illuminating gas, 6.68, 71
  from kerosene, 6.68, 71
  from stearin, 6.65, 66, 68, 71, 78
  hardening, 6.66, 68, 69, 70
  history, 6.65, 66
  manufacture, 6.68–70
  negative, 6.66, 77
  positive, 6.66, 78
  preparation, 6.66, 70, 71, 77, 78, 82
  quarter wavelength, 6.77
  soot-dielectric, 6.77, 78, 79, 82
SOR *See* Successive over-relaxation technique
Sorption phenomena, mirror electron microscopy and, 4.227
Sound field, 84.319
Source brightness, 11.111
Source code, GRASP library, 94.63–78
Sourceless Maxwell equations, 92.135, 159
Space:
  for matrices, 84.69
  paradigm, 94.261
  for templates, 84.91
Space charge, 1.225, 267
Space charge aberration, 1.267
Space charge effect, 83.49
  electron guns, in, 8.199, 203
  electron transport devices, 89.120, 130–134

Space mission images:
  Mariner, 4.92, 93, 94, 96
  Ranger, 4.88, 90, 91, 95, 97, 98, 99
  Surveyor, 4.104, 105
Spacetime algebra:
  classical and semiclassical mechanics, 95.374–377
  Grassman algebra, 95.377–379
  introduction, 95.273–278
  multiparticle quantum theory, 95.347, 349, 351
    applications, 95.379
    eight-dimensional streamlines and Pauli exclusion, 95.369–372, 374
    multiparticle wave equations, 95.364–366
    notation, 95.351
    Pauli principle, 95.366–369
    relativistic two-particle states, 95.361–364
    two-particle Pauli states, 95.352–360
  multivectors, 95.277
  operators, 95.276
  product types, 95.275
  reversion operation, 95.276
  rotors, 95.276
  spacetime calculus, 95.280–283
  spacetime split, 95.278–280
  spinors See Spinor theory
  vector derivative, 95.280–282
Space-variant image restoration, 99.293–294, 320–321
  image model, 99.300–309
  image restoration, 99.308–317
  image signal, 99.294–295
  Kalman filtering, 99.292, 293, 295
    estimation algorithm, 99.297–298
    state-space representation, 99.295–297
    steady-state solution, 99.299
    numerical results, 99.318–320
Spatial coherence, 7.158; 11.16
Spatial compounding, 84.341, 342
Spatial data structures, 86.147
Spatial filtering, 8.1–22

amplitude operation, 8.15–19
amplitude and phase operation, 8.20–22
complex filters, 8.20–22
defined, 8.2
filter design, 8.2, 4, 5
lens arrangement, 8.2
mask with maxima alignment, 8.17
methodology, 8.2–5
optical diffraction analysis by, 7.62–65
photochromic glass, 8.17–19
Schlieren systems, 8.11–14
substrate examined:
  analytical description, 8.7–10
  qualitative description, 8.5–7
Spatial frequency, 12.31
Spatial frequency filtering, image plate, 99.285
Spatial injection modulation of electrons (SIMTRON), 83.78, 79
Spatial partial coherence, 89.26–31
Spatial proximity, 86.147
Spatial reasoning, 86.114, 143, 148, 160, 162–164
Spatial resolution, 1.87; 12.321
Spatial sampling, visual cortex, 97.47–50
Spatial selection, 85.147, 215
Specialization order, 84.201
Special relativity, information approach, 90.147–149, 186
Specification:
  by abstract command, 86.99, 101, 103
  by example, 86.99, 105, 107
  through conversation, 86.99
  goal, 86.105, 107, 109, 112, 116, 120, 124
Specimen:
  damage, 2.196
    in electron microscope, 3.197, 200
    and high resolution, 4.87
    in scanning microscopy, 10.73–74
    variation in height, 10.30
  drift, image resolution and, 7.262
  ionization of, 1.157
  perturbation, and reflected electron beam phase, 4.233–237

INDEX 175

preparation, 83.212, 232, 242
  coating, 83.219, 233
  contamination, 83.239
  critical point drying (CPD), 83.239, 242, 245
  cryo-preparation, 83.252–255
  double-layer coating, Pt–C, 83.247–248
  fixation, 83.232–233
  freeze-drying, 83.232, 239
  freeze-fracture/thaw fix, SEM, 83.242, 245
  living, SEM, 83.209
  nuclear pore complex, 83.242, 245–246
  phase determination and, 7.191
  phase information and, 7.273
  phase problems and, 7.257
  with repeating units, phase problems and, 7.258, 259
  size, phase problems and, 7.246
  structure, phase information and, 7.256
  temperature, 1.152, 156
  thickness:
    phase error, 7.267
    phase information and, 7.273
    phase object approximation and, 7.194, 265
    stained, phase determination and, 7.269, 270
    transfer functions and, 7.119–122
Specimen holders, 11.76
Specimen stage, 11.8, 74, 76, 77
  cooling, 1.164
  design, 1.151
  mirror electron microscopy, 4.211–212
  space cooling, 1.159
Speckle, 84.326, 328
  attenuation effect, 84.329
  autocorrelation function, 84.331
  full-width-at-half maximum, 84.331
  fully developed, 84.330
  object-dependent, 10.72
  reduction, 84.342
  size, axial, lateral, 84.329, 331, 334, 338

Speckle interferometry, 93.143–144
Speckle noise, 92.13–14
Speckle patterns, 7.162, 172
Spectral element method, 82.361
Spectral methods, 82.360
Spectral signal-to-noise ratio (SNR):
  bright-field imaging, 94.220
  off-axis holography:
    STEM hologram, 94.246–252, 253
    TEM hologram, 94.218–221
  transmission electron microscopy, 94.211–213
Spectre II system, 5.45
Spectrofluorimetry, 8.59
Spectrogram:
  complex spectrogram, 97.9–10
  discrete spectrogram, 97.11
  reconstructing signal from, 97.19
Spectrometer constant, 12.182
Spectrometry, 11.120; 89.425–430
Spectromicroscopy, 10.55–58
  deep level transient, 10.67
  image storage, 10.55
  photoinduced transient, 10.67
  secondary ion, 10.303–304, 305
Spectrophotometry, 8.59
Spectroscopy, resonant tunneling electron spectroscopy, 89.135–136
Spectrum analyzers, 8.37
Spectrum-plate measurements, 8.32
Specular reflector, 84.333
Speech, 88.226
  autocorrelation function of, 82.104
  coding, 82.97
    linear prediction in, 82.120
    low-delay See Low-delay speech coding
    low-rate, 82.97
    performance criteria in, 82.100
    predictive, 82.150
  power spectral density of, 82.107
  probability density function of, 82.108
  production model, 82.102
  signal characterization, 82.103
  spectral flatness of, 82.109

Speed, propagation, 84.321
Spermatozoa:
  guinea pig, stripe photomicrograph, 7.98
  hamster, stripe microkymography, 7.98
  movement, microkymography and, 7.99
  movement registration, 7.90
  rhythmic motion curve, 7.90
  velocity, tracking microscope and, 7.11
Sphere, representation of, 13.29–30
Sphere-on-orthogonal-cone (SOC) field electron emission model, 8.223–225
  equipotential lines, 8.225
  lens aberrations, 8.241
Sphere packing, 97.216
Spherical aberration, 1.93, 129, 207–215, 216; 2.171, 210; 4.86; 5.165, 167, 173, 175, 193, 196, 197, 198, 244, 298; 93.207–218
  of asymmetric electrostatic lenses, 1.221
  axial astigmatism, 13.278–279
  of combinations of lenses, 1.219
  correction, 1.208
    adjustment, 1.268
    Archard's lenses, 1.239
    astigmatic systems, 1.232
    Burfoot's lens, 1.237
    coaxial lenses, 1.230
    combination of lens and mirror, 1.229
    Deltrap's system, 1.261, 266
    Glaser's corrector, 1.240
    grid lenses, 1.228
    high frequency lenses, 1.224
    improvement parameter, 1.265
    induced charges, 1.227
    prospects for, 1.269
    Scherzer's system, 1.233
    Seeliger's system, 1.236
    space charge, 1.225
    Whitmer's lens, 1.237
    zone plate, 1.270
  definition of coefficients, 1.205
  electron lens, 83.207, 229
  in electron microscopy, 7.287; 91.259–260, 267, 270, 277, 282–283
  electron optical transfer function, 13.274–279, 280–284
  in electron optics, 1.204; 11.38–43; 83.207
  fifth-order, 1.206, 236, 251, 263, 265
  Fraunhofer diffraction and, 13.261–268
  general expressions for the coefficients, 1.206
  lenses with minimum, 1.209
  in microscope objective, 14.258–259
  mirror-bank energy analyzers, 89.404–408
  monochromatic, 14.282–288
  multislice approach to lens analysis, 93.207–215
  scanning transmission electron microscopy, 93.86–87
  space-charge correction, 91.274
  thin film stack pattern, 14.96–97
  of thin helical lenses, 14.62–66
Spherical aberration constant, electron lenses, 1.122, 127, 132, 145, 205
Spherical capacitors, tests of, 13.45–47
Spherical cathode field electron emission model, 8.222
Spherical cathodes:
  brightness curves, different orientations, in, 8.185
  emission images, 8.183
Spherical mean, orientation analysis, 93.284
Spherical monogenic derivatization, 95.380–383
Spherical monogenic Pauli spinor, 95.303–305
Spherical wave, propagation in lens field, 93.194–195
Sphericity, 3.36, 64
Spheroidal graphite iron, 5.141, 142, 144
SPIDER (software), 86.98, 100, 102, 104
Spin, quantum mechanics, 90.163–164

Spindt cathodes, 83.15, 36–38, 47, 52, 55, 58, 76, 77
Spin dynamics, 97.337
Spin measurement:
  Dirac current, 95.339, 342
  relativistic model, 95.342–344
  spacetime algebra, 95.339, 342
  wavepacket simulations, 95.344, 346–347
Spinor theory, 95.283–284; *See also* Dirac spinor; Pauli spinor
  charged-particle wave optics, 97.258
  axially symmetric magnetic lenses, 97.333–335
  free propagation, 97.330–332
  general formalism, 97.322–330
  magnetic quadrupole lenses, 97.336
Spin–phonon coupling, 89.241–242
Spin polarization, single-electron logic devices, 89.223–235, 241–243
Spin-polarized low-energy electron microscopy (SPLEEM), magnetic microstructure, 98.333
Spin-polarized scanning electron microscopy, Japanese contributions, 96.716–718
Spin-polarized scanning tunneling microscope (SPSTM), 89.225, 236–237
Spin precession devices, 89.106, 199–203
Spin-spin coupling, quantum devices, 89.225–226
Split-brain research, 94.261, 263
Split detector microscopy, 10.50
Split gate structure, 91.214, 220, 227
Splitting plane, 85.151, 210
Splitting polyhedra, 85.152
Spoke diagram, 7.62
Sport size, 83.122, 181–185
SPSTM *See* Spin-polarized scanning tunneling microscope
Sputnik, 91.198
Sputtering, 83.13, 14
Sputtering coefficient for ions, 83.63
Sputtering damage, 83.53
Sputtering erosion, 83.33, 35, 62
Sputtering yield, 11.114
SQNR *See* Signal-to-quantization noise ratio
Square matrix, minimax algebra, 90.116, 117
Square pixels, 93.229, 233, 274
Square template, 90.362
SQUID *See* Superconducting quantum interference device
SRAM *See* Static RAM
S-SEED *See* Symmetric self-electro-optic effect device
STA *See* Spacetime algebra
Stabilization:
  high voltage, 2.190, 216
  lens current, 2.191, 220
Stack-based algorithm, 90.405, 417
Stacked capacitor cell, 86.65
Stacking faults, 2.233, 385
Stage:
  mechanical, 3.65
  point-counting, 3.75–77
Staining, 2.253, 2.268, 2.276, 2.325 et seq.
  contrast in electron microscopy of biological macromolecules and, 7.287
  image interpretation and, 7.186
  phase information and, 7.273
  weak phase approximation and, 7.266
Staircase avalanche photodiodes, 99.94–95
Standard scattering equation, 86.205
Standard X-ray area scan:
  advantages, 6.288
  background level, 6.289, 290
  corrections, 6.289
  CRT pulse reproduction, 6.291
  flexibility, 6.288
  gray level response, 6.289, 290, 291, 292
  gray scale, 6.291
  image definition, 6.290
  limitations, 6.289, 291

Standard X-ray area scan, *continued*
  orthographic projection of specimen, 6.288
  pulse superposition, 6.290
  scan speed variation, 6.288
  simplicity, 6.289, 291
  variables, 6.288
Staphylococci, 5.335; 6.160; 8.18
State functions:
  elementary state functions, 94.329–333
  nonelementary state functions, 94.333–335
  signal description, 94.324, 329
States:
  deep in the tail, 82.236
  estimation, 85.20, 25
  localized, 82.232
  variable, 85.21, 24, 39, 45, 48, 51, 53
States of information, regularization, 97.98–99
State-space, Kalman filtering, 99.295–297
Static current field, 82.4, 8
Static mass analyzers, 89.472–476
Static RAM (SRAM), 86.2
Static semiconductor permittivity, 92.116
Static SIMS, 11.104
Stationary orbit, satellite communications, 91.198–199
Stationary-phase approximation, 86.194, 195
Statistical analysis, orientation analysis data, 93.278–294
Statistical information theory, 91.37–41
  unified r,s-mutual information measures, 91.110–132
STD *See* Subscriber trunk dialing
Steady state:
  convergence to, 90.75–79
  minimax algebra, 90.58–74
  without strong connectivity, 90.70–74
Steel, thermal cycling, 10.297
Steel inclusions, and image analyzing computer, 4.364–368
Stellar spreckle interferometry, 93.143–144

STEM *See* Scanning transmission electron microscopy
Step edge, 93.231
Step microkymography, 7.76
  picture content, 7.89
  registration principles and techniques, 7.78
Step photomicrography, 7.77
Stereo, 85.114
  matching, 86.140
  vision, 86.86, 125, 139–140, 142
Stereo imaging, 11.154; 91.253–255
Stereological data:
  distribution, 5.134, 135
  estimation formula, 5.132, 133
  for human chromosome, 5.159, 160
  mean values, 5.133, 134
  for spheroidal graphite iron, 5.141, 142
  standard deviation, 5.132
  for vascular bundle cells, 5.140
Stereology, 5.115; 7.48, 65
Stereometry, 5.115
  area-measurement techniques, 5.120
  contiguity, 5.142
  counting criteria, 5.137
  directional factors, 5.143
  form factors, 5.144, 145
  linear analyses, 5.120, 121
  measurement of area, 5.140, 141, 142
  number of intersections, 5.138, 139, 140
  number of particles, 5.137, 138
  particle populations, 5.145–159
  point-counting method, 5.120
  problems, 5.116–120
  sample representativity, 5.135, 136
  statistical evaluation of data, 5.132
  verification of data, 5.132–135
Stereoscopic microscopy, maximum numerical aperture, 10.47–49
Stereo vision, 97.74–75
Stern–Gerlach apparatus, spin polarization, 95.287, 339
STFT *See* Short-time Fourier transform
Stigmatic focusing, 11.114
Stigmator, 1.154, 264

Still image coding, 97.226–232
Stimulated emission, 14.122
STL *See* Scanning tip lithography
STM *See* Scanning tunneling microscopy
Stochastical error, 85.15
Stochastic gradient, 82.132
Stochastic integration, image regularization, 97.119
Stochastic process, 85.17, 19
Stokes' and anti-Stokes lines, 82.265
Stoke's theorem, 13.175; 14.125; 82.16, 25
Storage capacitors, p-n junction, 86.6–32
Storage technology, 91.237–238
Storage time, 86.16, 20, 23, 25
Storage tube, 91.237, 238
Straight-edge, image, various systems, 10.22, 23–24
Strained-layer superlattice (SLS), 90.238
Strapdown navigation system, 85.8–9, 11
Stray ac fields, 11.12
STRAYFIELD (software), 13.43, 104, 113, 115
Stray fields, 83.221
Streak image microkymography, 7.76
Streak images, 1.17, 21, 32, 33, 36
Streak suppression, 92.19
Strehl intensity, 7.139
Strict cuts, fuzzy set theory, 89.264–265
Strict phase contrast, 6.90
Strioscopy, 5.300
Stripe detector, 94.237–239, 248
Stripe microkymography, 7.75–77
  blood flow determination, 7.96
  picture content, 7.89
  registration principles and techniques, 7.78
  single-slit diaphragm, 7.75
Stripe photomicrography, 7.74, 88
  equipment for, 7.84, 85
  exposure time determination, 7.87
  pancreas section, 7.95
  picture content, 7.89–94
  principles, 7.81–82
  procedure, 7.86–89
  quantitative evaluation, 7.89–94
  recording technique, 7.86–89
  registration principles and techniques, 7.78
Strip processor, 99.314
Strip surface currents, 92.106–108
Stroboscopic measurements, 12.144, 146
Stroboscopic mirror electron microscopy, 4.249, 256, 257
Stroboscopic technique for flow velocity, 1.8, 22
Strong connectivity, 90.48–49
Strong exchange interactions, ferromagnets, 98.325
Strongly definable partition, 94.168
Strong realization problem, DES, 90.108–109
Strong transitive closure, 90.45–46
Strowger switch, 91.192, 202
Structural properties method, 94.316
  applications:
    data compression, 94.386–391
    extrapolation of time series and real-time processes, 94.391–392
    filtering and rejection of signals, 94.379–383
    malfunction diagnosis, 94.325, 392–394
    parameter estimation, 94.365–376
    signal identification, 94.383–386
    signal separation, 94.376–379
  definitions, 94.327–329
  elementary state functions, 94.329–333
  generating differential equations, 94.324, 327, 336–347
  nonelementary state functions, 94.333–335
Structural resolving power, 7.138, 140
Structure conservation, biological macromolecules, 8.114–119, 122–124
Structure constant, 84.277–278
Structured centrosymmetric matrix:
  autoregressive parameter estimation, 84.290–291, 293–296, 301

Structured centrosymmetric
matrix, *continued*
estimation performance:
autoregressive parameters,
84.292–293, 302–309
covariance, 84.287
ideal structure, 84.278–279
isomorphic block diagonal form,
84.280–281
isomorphism of simple algebras,
84.276–278
Jordan subalgebra dimension,
84.275–276
relation to Toeplitz matrix, 84.270
structure set, 84.272–275
trace covariance bound, 84.285–286
Structured covariance estimation, abstract
algebra and, 84.262–313
Structure factor, 86.189
Structure refinement:
density flattening, 88.146
effect of dynamical scattering, 88.150
Fourier refinement, 88.145
least squares refinement, 88.143
Structure set:
commutative, 84.271–272
for Dirac matrix, 84.277
extension:
quotient space, 84.272–273
recursive, 84.274–275
free, 84.272
inverse covariance, 84.272, 313
for minimum variance estimation,
84.286
symmetric centrosymmetric matrix,
84.272–275
Toeplitz covariance matrix, 84.269–270
Structuring element, 84.65, 87–88
Structuring functions, 89.368; 99.9, 13–15,
18–20
convex structuring functions, 99.13–14
dimensionality, 99.40–42
flat structuring function, 99.44
poweroid structuring functions,
99.14–15, 41, 42–46

quadratic structuring functions, 99.15,
45–46
scale-space, 99.37–46
semi-group properties, 99.15, 37–38
signal extrema, 99.20–22
Stub-tuning, 89.179; 91.226
Styryl dyes, fluorescent probe, 14.156–158
Sub-band coding, 82.158
Subcomplex:
defined, 84.205
mapping, 84.236
Subcomposition, 89.271, 283–284
Subgraph isomorphism, 84.229–235
Subneuron field, 94.305–309
Suboptimal algorithms, 97.124–127
Subscriber trunk dialing (STD), 91.192
Substrates:
aluminum, 83.34
microscopic examination, 8.7–10
qualitative description, 8.5–7
refractive index, 8.6
sapphire, 83.36
scattering, 90.241, 243–245
silicon, 83.29
surface, 87.97
transparent, 8.6
Subthreshold current, 86.37
Subtraction:
Minkowski subtraction, 89.328
optical symbolic substitution, 89.60, 61,
64–65, 70
Subtractive display, flat panel technology,
91.246–247
Successive approximation quantization:
convergence, 97.211–214
orientation codebook, 97.214–216
scalar case, 97.205–206
vectors, 97.207–211
Successive approximation wavelet lattice
vector quantization (SA-W-LVQ),
97.191–194, 252
coding algorithm, 97.220–226
image coding, 97.226–232
theory, 97.193–220
successive approximation

quantization, 97.205–220
wavelet transforms, 97.195–205
video coding, 97.232–252
Successive line over-relaxation (SLOR) methods, 13.157, 159
Successive over-relaxation (SOR) technique:
    coupled linear equation solution by, 8.227
    modeling electron optical systems, 13.6
    as numerical method, 13.157, 159
Succinic dehydrogenase:
    distribution, 6.216
    ferricyanide technique, 6.216, 222
    localization, 6.214
    mitochondria, 6.214, 216, 220
    tetrazolium salt technique, 6.214
Sudden perturbed approximation, 86.207, 208, 210
Sulfatase, 6.181
    electron microscopy, 6.181, 182, 183
    histochemistry, 6.179, 180, 221
Sum of deviation functions, 86.205
Supercomposition, 89.271, 283–284
Superconducting electron lens, 10.238–241
    aberrations, 5.211, 214, 216, 217, 234
    assessment, 5.234
    astigmatism, 5.202, 217
    cryogenic equipment, 5.211, 225, 235
    diamagnetic flux shield type, 5.228–233
    for electron microscopes, 14.24–25
    flux density distribution, 5.209, 210, 212, 213, 214, 216, 217, 218, 220, 221, 225, 226, 227, 229, 230, 232
    Glaser function, 5.212, 213, 214
    with iron pole piece, 5.224, 225
    performance, 5.235
    resolution, 5.202, 211, 234, 235
    Glaser model and, 5.225, 226, 227
    iron-shrouded solenoid, 5.218–220
    microscopes with, 14.23–24
    with pole pieces, 5.224–233
    with rare earth pole piece, 5.225–228
    resolution, 5.234
    simple solenoid, 5.215–217
    trapped flux type, 5.220–224
    without pole pieces, 5.215–224
Superconducting quantum interference device (SQUID), 89.98
Superconductors, 5.201
    in electron microscopy, 14.16–25
        diamagnetic shielding lenses, 14.20–21
        early work, 14.16–17
    lens, 5.201, 202
        shrouded coils and pole piece lenses, 14.21–23
        solenoid lenses, 14.18–19
        trapped-flux lenses, 14.19–20
    filamentary, 5.207
    Ginzberg–Landau theory, 14.9–11
    high field, 5.205–208
    high-temperature *See* High-temperature superconductivity
    layered structure of, 14.29–30
    London theory, 14.7–9
    magnetic flux structures, 14.13–15
    magnetization curves, 5.203, 206
    microscope theory, 14.11–13
    mirror electron microscopy, 94.98
    properties, 14.6–7
    scanning susceptibility microscopy application, 87.184
    two-fluid model, 14.8–9
    type I, 5.202, 203, 205
    type II, 5.203, 205–208, 240; 87.169–176
        commercial, 5.208
        constant field magnitude contours, 87.174–175
        critical current density, 5.208
        critical field, 5.208
        critical temperature, 5.208
        domain walls, 5.240
        flux jumps, 5.207
        Ginzberg–Landau equations, 87.172–173
        instabilities, 5.207
        magnetic configurations, 5.240
        magnetization curve, 5.206

Superconductors, *continued*
  magnetostatic boundary value problem, 87.169–170
  pinning points, 5.206
  reciprocal lattice, 87.171
  resistive effect, 5.206
  steady state properties, 5.205, 206
  surface stray field, 87.172–173
  vortex lattice deformation, 87.175–176
Superimposed lattices, crystal-aperture scanning transmission electron microscopy, 93.96
Superlattice reflections, 11.89
Superparamagnetic probe, 87.182–183
Superquadrics, 86.149–150
Supply-function, field electron emissions, 8.211–213
Support films:
  for electron microscopy, 8.110
    charge, 8.129
    glow discharge treatment, 8.120, 127–130
    hydrophilic properties, methods of producing, 8.120–121, 128–129
    preparation, 8.119–121
    properties, 8.111–113, 129, 132
  for microincineration, 3.105, 134
  structure of, 1.171
Surface, 83.7, 10, 17, 28, 34, 35
  atomic steps, 11.64
  decoration, 11.58
  dynamic processes, 11.91
  electrostatic potential, 87.106–107
  imaging, 11.57, 58
  ionization sources, 11.106
  models, 85.96
  netted, 85.191
  opaque, 85.113
  rendering, 85.133
  structure, 2.401
  substrate surface, 87.97
  treatment, 86.39
Surface brightness, field electron emissions, 8.216, 248–251

Surface collapse, during particle adsorption, 8.16
Surface conductivity, mirror electron microscopy, 4.226
Surface current, density, 82.26–32, 41
  coefficients of, 82.2–3, 17–18, 28–35
  horizontal, 82.3–8, 14–15, 17–18, 26–27
  vertical, 82.2–3, 8–10, 14–15, 26–27
Surface effects, 87.217
Surface generation velocity, 86.14, 29
Surface potential, 12.163
  mirror electron microscopy, 94.98
Surface resonance, 90.317–334
Surface resonance scattering, RHEED, 90.323–334
Surface spreading technique, chromosome superstructure study, in, 8.93–97
Surface states, 87.228
Surface tension, 8.111–112
  damage to biological macromolecules, 8.115–116
Surface/volume ratio, 3.77–80
Surface waves, 82.7–8
Susceptibility meter, 91.174, 175
SVD *See* Singular value decomposition
Swing objective immersion lens (SOIL), 83.170–181
Swing objective lens (SOL), 86.162, 173
Switching speed, single-electron cells, 89.241–242
Switching systems, 91.164, 192, 201–204
  computer-controlled, 91.164, 201–204
  electromechanical, 91.192, 201–202
  electronic, 91.202
  exchange switch systems, 91.201–204
  opto-electronic, 91.209
  photonic, 91.209
  plasma, 91.250
  quantum-coupled devices, 89.221, 222, 225, 241–242
  Strowger switch, 91.192, 202
SYCOM satellite, 91.199
Symbolic substitution, optical *See* Optical symbolic substitution

INDEX

Symmetric self-electro-optic effect device (S-SEED), 89.79–80
Symmetries, group of, 82.357
Symmetrized Chernoff measure, 91.126–129
Symmetry properties, projections, 7.307–310
Synapse, 86.74
Synchronization, hardware support, 87.272–273
Synchronous model of parallel image processing, 87.292–293
Synchrotron radiation, 10.323
Synthetic aperture radar (SAR):
  detail enhancement, 92.45, 46
  extremum sharpening, 92.33, 46
  max/min-median filter, 92.41
  smoothing of speckle noise, 92.13–14
  top-hat transformation, 92.64
Synthetic scanning, 8.43–44
System clock, 91.233–234
System matrix, 90.4, 14, 79
System model matrix, 85.59, 64
System noise matrix, 85.24–25, 36, 38, 47, 57, 64–65
System state vector, 85.17, 36, 41–43, 45, 47, 50, 52
Systolic array computer, 85.25–26

T

TAAC-1, 85.172–175, 177
T-layer structural preservation, 8.115
Tachometer, 1.20
Takahashi, Kanjiro, 96.660
Tangential component:
  of current vector potential, 82.19, 28, 35
  of electric field intensity, 82.5, 19, 22, 26, 28, 36, 39, 81
  of magnetic field intensity, 82.7, 8, 15, 22, 28, 29, 32, 33, 35, 36, 39, 70, 80
  of magnetic vector potential, 82.16, 17, 29, 36
Tangent vector, 84.137, 171, 173, 189

Taste, 85.81
TAT 8 system, 91.200
τ-mapping, 89.330–332, 383
  anti-extensive, 89.350–351
  binary, 89.336, 366
  cascaded, 89.345–348, 363–366, 381–383
  dual, 89.348–349, 354–358
  extensive, 89.350–351
  gray-scale, 89.336–337, 366–374
  intersection, 89.339–343
    reversing, 89.376–377
  overfiltering, 89.351–353
  translation, 89.337–338
  underfiltering, 89.351, 353–354
  union, 89.338–339, 342–343
    reversing, 89.375–376
Taylor series expansion, in FDM methods, 13.7, 12
TCQ *See* Trellis-coded scalar quantizers
T-divergence, 91.40, 79–82
TDS scattering, 90.286–287
TEAM workshop, 82.75, 80
Technical drawings, 84.254–257
Technologicality, 94.325
TED *See* Total energy distribution
Tektronics, 91.234, 237, 241, 250
Telecentric optical system, microscope objective, 14.329–332
Telecentric slit objectives, 8.43
Telecommunications, 91.151, 164
  broadcasting, 91.207–209
  coaxial cable, 91.193–194
  computer-controlled switching, 91.164, 201–204
  digital revolution, 91.195–197
  electronic newspaper, 91.209
  facsimile machine, 91.204
  future, 91.205–212
  guided missiles, 91.149, 174
  history, 91.190–205
  microchip, 91.151, 195, 202
  microwave radio relay, 91.193–194
  optical fiber communications, 91.199–207

Telecommunications, *continued*
  analog, 99.68
  history, 2.12, 13; 91.179–180, 182, 199–201
  satellite communications, 91.198–199
  teleconferencing, 91.150, 189–190, 205, 206, 208, 211
  teletext, 91.204–205, 241
  transistors, 91.195
  United Kingdom, 91.191–194, 196–205
  video library service, 91.207–208
  virtual reality, 91.208–209
Teleconferencing, 91.150, 189–190, 205, 206, 208, 211
Telefocus electron gun, 10.247
Telescope, radioastronomy, 91.285–290
Teletext, 91.204–205, 241
Television, 91.207–209
  flat screen, 91.236
  frame store, 91.242
  history, 91.193–194, 196, 198
  image intensifier design, 9.29–46
    camera tubes, 9.39–44
    channel plates, 9.32–33
    charge-coupled devices and, 9.31, 44, 45
    current density dependence of DQE, 9.31
    directly bombarded Se-target, 9.32
    gain, normalization mode and, 9.31
    image intensifier tubes, 9.38–39
    KCl-target, 9.32
    "quantum level diagram," 9.30–31
    rms noise of video amplifier, 9.31
    Si-mosaic target, 9.32
    solid-state image converters, 9.44–46
    transmission fluorescent screen, 9.31, 33–38
  plasma display tube, 91.245–246
  three-dimensional, 91.253, 254
  United Kingdom, 91.193
  video recording system in microkymography, 7.93
Television cameras, 4.145–148

Television microscope, 5.129–132, 137, 142, 156
  advantages, 5.129
  electronic slide-in units, 5.131
Tellurium, properties, 92.82
TELSTAR, 91.198, 199
TEM *See* Transmission scanning microscopy
Temperature:
  bandgap narrowing and, 82.270
  critical temperature, 83.13
  electron bombardment effect on specimen, 96.258, 260–261
Temperature-field domains, electron emission mechanisms, for, 8.209, 213, 214
Template:
  additive and multiplicative conjugates, 84.78
  constant, 84.82
  correspondence with structuring element, 84.85
  decomposition, 84.78, 84
  defined, 84.76
  induced functions, 84.82
  null, negative and positive, 84.82
  one-point, 84.82
  operations between image and template, 84.78
    backward and forward additive maximum, 84.80
    backward and forward linear convolution, 84.79
    backward and forward multiplicative maximum, 84.80
    continuous domain, 84.81
    generalized backward and forward, 84.79
    multiplicative additive and minimum, 84.81
  operations between templates, 84.81
    convolution type, 84.83, 84
    pointwise, 84.82
  row/column/doubly-P-astic, 84.98
  strictly doubly F-astic, 84.110

support, infinite negative and positive, 84.77
target point, 84.77
translation invariant and variant, 84.78
transpose, 84.78
Temporal redundancy, 97.235
Tensor:
  permeability, 82.3, 39, 41, 42, 61
  permittivity, 82.3, 39, 61
Tensor products, group algebra, 94.16–17
Tensor RHEED, 90.276–279
Tenth-order Hamiltonian function, power-series expansion, 91.11–13
Terminating index, 90.81
Ternary signed-digit number system, optical symbolic substitution, 89.59
Tesla, commercial microscope production, 96.452, 474, 495, 503, 506, 512, 517–518, 532–534
Tessalation, surface, 85.108
Test objects, for contrast transfer properties, 4.79–80
Test signal, malfunction diagnosis by structural properties method, 94.325, 392–394
Tetracene crystals, 5.310, 316
Tetrazolium salt technique, 6.212–215
T-even bacterial virus, 6.232, 241
Texas Instruments:
  quantum-coupled integrated circuits, 89.218, 221–222
  semiconductor history, 91.143, 147, 149, 169, 195
Texture:
  analysis, 84.329
    Gaussian wavelets, 97.64–68
  defined, 95.387–388
  feature image, 95.390–391
  generation, 84.325
  nonparametric analysis, 84.337, 345
    cooccurrence matrix, 84.337
    MAX–MIN method, 84.338
  power spectrum, 84.336
  representation:
    cooccurrence matrix approach, 95.389

  feature frequency matrix, 95.388, 390–391, 393, 400–403, 405–406
  image transformation, 95.389
  local mask methods, 95.389
  multichannel filter, 95.389
  properties, 95.390
TGZ 3 Particle Size Analyzer, 5.154, 155, 156, 157, 159
THEED *See* Transmission high-energy electron diffraction
Theorem of Gondran and Minoux, 90.115–117
Thermal diffuse scattering, 86.186
Thermal drift, 1.150, 164
Thermal fading, photochromic glass, 8.18
Thermal-field emission guns, 9.191
Thermal waves, 10.58; 12.313, 317
Thermionic cathodes, 10.244–248
Thermionic emission, 83.26–29; 86.53; 87.218
  angular emission distribution, 8.215
  current density, 8.213
  lens action, analytical model, 8.223
Thermionic-field emission, 86.53
Thermodynamics, information approach, 90.189
Thesaurus construction, fuzzy relations, 89.307, 310–311
Thick lens theory, 93.192–194
Thickness, measurement with electron off-axis holography, 89.37–38
Thin-film electroluminescence, 91.245
Thin film electronic devices, imaging, 4.248
Thin film lens elements:
  concept of, 14.66–69
  introduction to, 14.66–73
  magnetic field of, 14.73–80
    and Biot–Savart law, 14.73–75
    multipole expansion, 14.77–80
    of one-turn spiral, 14.75–77
  types of, 14.69–73
Thin-film probes, 87.148–157
Thin films:
  French contributions, 96.110–112, 120

INDEX

Thin films, *continued*
  Japanese contributions, 96.611, 614
  magnetic, 98.387–399, 400–401, 402–408
  mirror electron microscopy, 4.226
  Southern Africa contributions, 96.336–337
Thin film stack pattern:
  capacitive alignment, 14.81–83
    basic principles, 14.81–82
    differential capacitive position sensor, 14.82–83
  geometry, 14.83–90
    adjacent layers, connection between, 14.88–90
    capacitive electrodes, 14.85–88
    helical pattern, 14.84–85
    introduction, 14.83–84
  optical properties, 14.90–99
    field distribution, 14.92–94
    first-order properties, 14.94–95
    introduction, 14.90–92
    parasitic aberration 14.97–99
    spherical and chromatic aberration, 14.96–97
  prototype element, experimental work on, 14.100–110
Thin film superconductors:
  analysis of, 14.51–53
  critical current density in, 14.36–38
  evaporation techniques, 14.47–53
  experiments with, 14.54–58
  fabrication, 14.40–54
    equipment, 14.47–48
    evaporation techniques, 14.47–48
    laser ablation techniques, 14.45–47
    other techniques, 14.53–54
    procedure, 14.48–51
    sputtering techniques, 14.43–45
  field distribution, 14.56–57
  introduction to, 14.41–43
  measurements in, 14.55–57
  patterning, 14.54–55
  stack pattern, design of, 14.80–99
    capacitive alignment, 14.81–83

  geometry of, 14.83–90
  optical properties, 14.90–99
  superconducting properties, 14.55–56
  technical feasibility, 14.57–58
Thin helical lenses:
  focal properties, 14.58–62
  introduction to, 14.58–66
  spherical aberration, 14.62–66
Thinning, three-dimensional, 85.186
Thomas–Fermi approximation, 82.210
Thomas–Fermi model, space charge effects, 89.132–134
Thouless energy, 89.108
Thouless temperature, 89.107
Three-dimensional circulators *See* Circulators
Three-dimensional crystallographic group, group-invariant transform algorithms, 93.2–3
Three-dimensional display, 91.252–253
Three-dimensional images reconstruction, 7.321–324, 329, 330; 97.59–60
  mathematical apparatus, 7.304–330
  synthesis by projection functions, 7.314
  synthesis of modified projection functions, 7.318–321
Three-dimensional imaging, multiwave, with fluorescence, 14.184–189
Three-dimensional microscopy, 14.214–215
Three-dimensional orientation data, 93.283–287, 325
Three-dimensional systems
  charging effects on insulating specimens, 13.77–78
  computation, high-beam energy scan, 13.79–82
  focusing, in SEMs, 13.113–119
  object representation in, 13.26–30
  software for, 13.25–44
  spherical capacitor tests, 13.45–48
  testing software for, 13.44–59
Three-dimensional thinning, 85.186

Three-electrode mirrors, energy analyzers, 89.417–420, 458–469
310 field emissions, STEM, 93.58–59, 79–87
Three-port circulator, scattering parameters, 98.117–120, 238–245, 302
Threshold, choice of, 88.337
Thresholded posterior means (TPM) estimate, 97.88, 120
Thresholding, 88.301, 336; 93.223, 231–232
Threshold setting errors, image analyzing computer, 4.375–378
Through-focus, 1.171; 86.181, 182
Through lens detector, 12.93, 98, 108, 125, 131
Through-thickness, 86.181, 198
TICAS system, 5.45
Tilt, parasitic aberration and, 86.245, 250, 260
Tilted beam holography, 7.167
Tilted illumination, 11.24
  bright field microscopy, 7.249–251
  in STEM, 7.147–158
TI mapping *See* Translation-invariant set mapping
Time–energy relation, uncertainty principle, 90.167–168
Time of flight spectrometer, 11.143; 12.109
Time-gain-compensation, 84.329, 340, 344
Time-harmonic, 82.75
Time-measuring microscopes, 8.33
Time scripts, 94.290, 291
Time series, extrapolation, 94.391–392
Time variation, sinusoidal, 82.38, 40, 41
Tip cathode electron guns, analytical field calculation, 8.228
Tips; *See also* Microtips; Nanotips
  disruption of, 83.13, 21, 22, 63, 64
  field emission, 83.12–17, 33, 43, 47, 53, 58, 63, 76
  single crystal, 83.33
TIRF *See* Total internal reflection fluorescence

Tissue mimicking phantom, 84.340, 344
Titanium alloy, photoelectron microscopy, 10.298
TLEED *See* Transmission low-energy electron diffraction
TN cell *See* Twisted nematic cell
Tobacco mosaic virus, 6.227, 232
  electron micrograph, 6.271
  modulus of Fourier transform, 7.314
  negatively stained, structure and, 7.292
  polydiscs, 7.349, 350
  preparation, 6.263
  rods, 6.253, 254
  stacked disc protein, 6.243, 247
  stacked disc rods, 6.248
  three-dimensional structure, 7.347–349
  two-dimensional distribution, 7.313
Toboggan algorithm, 92.34
Toepler's schlieren method, 8.11
Toeplitz matrix:
  biased correlation estimate, 84.303
  estimation performance:
    autoregressive parameters, 84.292–293, 302–309
    covariance, 84.287
    SIR, 84.270–271
  inverse, 84.271–272
  lowest Jordan subalgebra dimension, 84.275–276
  maximum likelihood estimate, 84.269, 282, 285
  relation to symmetric centrosymmetric matrix, 84.270
  role in autoregressive parameter estimation, 84.289–291, 293–296
  structure set, 84.269–270
  trace covariance bound, 84.285–286
Tohoku University:
  electron microscope development, 96.245–248
  pointed cathode studies, 96.689–690
Tolansky multiple beam interference, 6.11!
Tomography; *See also* Computed tomography
  image formation, 97.59–60

Tomography, *continued*
  image reconstruction, 97.159–166
Tonicity, in fixation, 2.290, 302, 307, 310
Top contouring, 93.287
Top-down analysis, 86.160
Top-hat transformation, mathematical morphology, 92.63–64
Topographical imaging, 7.232
Topographical maps, 84.250–254
Topographic contrast, 83.206, 213, 218, 231, 235–236
  evolution, 83.220
  high resolution, 83.228, 236
Topological relations, knowledge representation in logic based on, 86.143–147, 164
Topological space, axioms, 84.201
Toroidal magnet, 86.267
Toshiba Corporation:
  EUL series, 96.676–677
  Toshiba No. 1, 96.673, 675
  Toshiba No. 2, 96.675–676
Total electron gun beam:
  energy distribution, 8.200–203
    across the beam away from the axis, 8.202
    half-width, 8.201, 202
    optic axis, on, 8.201–202
  space charge effects on cross-section, 8.199
Total energy distribution (TED), field electron emission, 8.214–215
  curves, 8.216, 219
  full width at half maximum (FWHM), 8.220
Total internal reflection fluorescence (TIRF), 14.178–179
Totally undefinable partition, 94.170–171
Total reflection angle X-ray spectroscopy, reflection high-energy electron diffraction (RHEED) and, 11.94
Total reflection fluorescence, 14.178–180
Total residual error, 84.12
Touch, 85.81
Touch-tone dialing, 91.202

TPM *See* Thresholded posterior means esstimate
Trace analysis, 11.113, 142
Trace inner product, 84.265, 270
Trachyte, 7.37
Tracking, boundary, 84.220
Tracking microscope, 7.1–15
Trajectory analysis, electron beams, 8.157–162
Trajectory equation:
  build-up process, 8.253
  field electron emission, 8.244
Transaxial mirrors, 89.392, 396
  charged particle focusing and energy separation, 89.433–441
  energy analyzers, 89.441–469
  mass analyzer, 89.469–476
Transconductance, 83.69
Transducers:
  aperture, 84.344
  array, 84.344
  backing medium, 84.319
  continuous wave, 84.319
  directivity function, 84.320
  dynamic focus, 84.344
  geometrical, 84.331
  linear array, 84.341
  multifocus mode, 84.344
  phased array, 84.341
  piezoelectric layer, 84.318
  pulsed mode, 84.319
  pulse-echo mode, 84.318
  pulse waveform, envelope, 84.319, 331
  synthetic focus, 84.345
  tracking microscope, 7.3, 4, 5, 12
Transfer functions, 7.117–122; 12.27, 31
  extended sources, 7.123
  theory, 7.102
  tissue, 84.320, 323
  for weakly scattering specimens, 7.122–140
Transfer lens system:
  advantages, 6.7
  focal length, 6.7
  image position, 6.6

magnification, 6.6, 7
resolution, 6.6, 7
Transfer matrix, 86.232, 234
Transfer process rules:
  LLVE, dependency, 86.121, 123, 138
  selection, 86.121, 123, 137–138
Transform algorithms, phase retrieval, 93.110
Transformation:
  geometric transformation, 85.182
  viewing transformation, 85.117
Transformation equation, 85.11–12
Transformation group, 84.133, 84.146, 84.184 ff.
Transformation operator, 92.102–103
Transform coding, 97.51, 193
  bit allocation, 88.4
  block diagram, 88.3
  quantization error, 88.4
Transform efficiency, 88.45
Transforms:
  C-matrix (CMT), 88.41
  cosine *See* Discrete cosine transform
  cross-grating Fourier transforms, 8.16
  distance transform, 92.72–75
  extended Cooley–Tukey fast Fourier transform, 93.47–52
  fill-in, 86.110
  Fourier *See* Fourier transform
  Fourier–Mellin transform, 84.159
  Gabor–DCT transform, 97.52
  Gabor transform, 87.34–37
  group algebra, 94.27
  Hartley transform, 93.139–143
  Hilbert transform, 7.230, 231, 237, 242, 243; 93.111–116, 120–121
  Hilbert transform pair, 88.308
  integer *See* Integer transform
  integral *See* Integral transform
  Karhunen–Loeve, 88.7
  Laplace transform, 85.16, 60
  Lapped orthogonal transform, 97.11
  lattice *See* Lattice transform
  LPCH transform, 84.145
  matrix transforms, 94.27–44

medial axis transform, 85.186
minimum transform, minimax algebra, 90.114, 116, 117
mixed time–frequency transforms, 94.320–321
optical *See* Optical transform
optimal, 88.5
orthogonal *See* Orthogonal
sequential decomposition, 86.110
short-time Fourier transforms, 94.320, 321
sine *See* Discrete sine transform
sinusoidal, 88.10–15, 25
  integer, 88.24–40
  integer, derivation, 88.30
split decomposition, 86.110–111
symmetry cosine, 88.9
Walsh, 88.15, 25; *See also* Walsh matrix
watershed transform, 99.46, 47–48
wavelet transforms *See* Wavelet transforms
Zak transform, 97.29–30
Transillumination, 6.4, 9, 11, 13
  condenser system, 6.21–22
  darkfield, 6.17, 18, 21
  darkfield–brightfield, 6.21
  fluorescence, 6.21, 24
  light pipe system, 6.26–28
  light sources, 6.22–26
Transimpedance amplifier, 99.73
Transistors, 83.67; 91.143–145
  Aharonov–Bohm effect-based devices, 89.145, 146, 189
  alloy transistor, 91.144
  bipolar transistors, 89.136
  diffused transistor, 91.145
  discrete transistor, 91.146
  electrochemical transistor, 91.144
  field-effect transistors *See* Field-effect transistors
  gallium arsenide, 91.180–181
  granular electron transistors, 89.205–208
  high electron mobility transistor, 91.186
  history, 91.142, 143, 178–188
  planar transistor, 91.145–146

Transistors, *continued*
  p-n-p alloy transistor, 91.144
  quantum interference transistors, 89.106, 157, 160–165, 167, 178
  resonant tunneling transistors, 89.136, 245
  spin precession transistors, 89.106, 199
  telecommunications, 91.195
  T-structure transistors, 89.178–193
Transition matrix, 85.18–19, 36, 47
Transitive closure matrix, 90.54–55, 79–80
Translation, 84.169, 177; 85.118
  autotasking, 85.294–297
  τ-mapping, 89.337–338
Translation-invariant set mapping (TI mapping), 89.358–366
  basis algorithms, 89.361–363
  general basis algorithm, 89.363–365
Transmission amplitude, 89.124–125
Transmission coefficients, 90.318
Transmission cross coefficient, 7.120
Transmission electron microscopy (TEM), 90.210, 211, 289; 91.260, 274, 280–282; 93.69; 94.197
  commercial development in Europe, 96.470–486, 494–499, 513–519, 639–641, 645–647
  crystal-aperture scanning transmission electron microscopy, 93.57–107
  defocusing method, 5.240, 241
  development, 95.4, 18, 20, 25, 37
  field emission gun, 96.402
  image formation, 93.174; 94.203–213
  imaging plate with, 99.262–263, 265–269, 288
  invention, 96.11–12, 65, 132–135
  low-dose imaging, 94.207–213
  microfabric analysis, 93.222, 223
  off-axis holography, 94.213–221, 252–253, 253–256
  resolution, 96.144, 402, 524, 526
  ultra-high voltage, 11.66
Transmission high-energy electron diffraction (THEED), 90.210, 236–237, 290, 317

  by deformed crystal, 90.238–241
  by multilayer system, 90.238
  sensor THEED, 90.255–256
Transmission low-energy electron diffraction (TLEED), 90.321
Transmission SAM, 11.155
Transmission scanning near-field optical microscopy (SNOM), 12.266
Transmitted beam amplitude, 90.229
Transparency, 85.138
  of description tool, 94.326
  optical, 1.182
Transparent network, 91.210
Transparent objects, refractive index, 8.6
Transport theory, quantum mechanics, 91.215–216
Transputer, 85.165
Transversal filters, 82.2–39
Transverse field formulas, three-dimensional microstrip circulators, 98.133–134, 170–174
Trapezoidal fuzzy quantity, 89.320
Trapped flux electron lens:
  advantages, 5.223
  astigmatism, 5.223, 224
  construction, 5.222, 223, 224
  disadvantages, 5.224
  electron microscopy, 14.19–20
  flux density, 5.220, 221
  flux jump, 5.221, 223
  pinning points, 5.224
  superconducting cylinder type, 5.221, 222
  superconducting discs type, 5.222, 223
Traps, 87.222, 225, 227, 235; 89.221–222
Traveling-wave tube amplifier, 91.194
Traversing image data, 88.94, 104
Tree coding, 82.184
Trellis-coded scalar quantizers (TCQ), 84.51
Trellis coding, 82.186
Trench capacitor cell, 86.65
Triangular compositions, applications, 89.297–311

Triangular conorm, fuzzy set theory, 89.261
Triangularization, 85.24, 37, 42, 47
Triangular norm, fuzzy set theory, 89.261, 278, 289
Triangulation, 85.109
Triboluminescence, 14.122
Trilinear interpolation, 85.127
Triode electron guns, 8.137–204
Triodes, 83.67–72
Tripropylamine, glow discharge in, 8.126
Trochoidal analyzer, 12.114, 115, 209, 211, 214
Tropomyosin, actin complex, reconstruction, 7.354
Trüb–Tauber Company:
  commercial microscope production, 96.453–454, 486, 488
  electron microscope, 10.226
True covariance matrix model:
  free parameters, 84.266
  inverse linear, 84.271–272
  linear, 84.266–268, 273, 282
  nonsymmetric, 84.264
  orthogonal complement identity, 84.268
  simple symmetry, 84.265–266
Truncated potential mode, 90.244–245
Truth table:
  minimization, 89.66–68
  NAND logic gate, 89.230–231
T-structure transistors, 89.178–193
  analog, 89.188–189
  digital applications, 89.189–191
  electro-optic applications, 89.191–193
Tsu–Esaki formula, 89.115, 136–142
Tubular crystals, globular proteins, three-dimensional structure, 7.339, 340
Tubules, 83.34
Tubulin, three-dimensional structure, 7.341, 347
Tungsten, crystal-aperture scanning transmission electron microscopy, 93.58–59, 79

Tungsten field emission cathodes, 8.252–253
  emission pattern, oxygen processing, during, 8.254
Tungsten filaments, 9.191–192
Tungsten hairpin electron gun, 10.248–249
Tungsten thermionic cathodes:
  emission current density, 8.179, 213
  emission images, 8.184
  temperature-field domains, 8.214
Tunnel diode, 89.98
Tunneling, 82.223; 83.10, 13, 20, 21, 23, 25, 61; 86.53
  bandgap semiconductor, 99.88
  Dirac current, 95.333, 336
  incoherent tunneling, 89.128
  quantum mechanical tunneling, 89.136–142, 216
  resonant tunneling, 89.121–142, 126, 128, 222, 245
  sequential resonant tunneling, 89.129
  spin-polarized scanning tunneling microscope, 89.225, 236–237
  time calculation, 95.326–327, 337–338
  two-dimensional simulation, 95.338–339
  wavepacket tunneling, 95.332–333, 336–338
Tunneling time, quantum mechanical, 89.136–142
Tunnel resonance effects, field electron emissions, in, 8.220
Turbulence, coherent imaging, 93.153–166
Turing test, machine intelligence, 94.266–272
Turnip yellow mosaic virus, 6.234
  concentrated preparation, 6.263
  electron micrograph enhancement, 6.243, 248
  reconstruction, 7.371, 373
  three-dimensional reconstruction, 7.367, 368, 370
12,2 formula, 93.252, 253
12,9 formula, 93.235, 237–239, 252, 253, 257, 261–266, 277

20,2 formula, 93.252, 253
20,5 formula, 93.252, 253, 257, 261–266, 269, 272, 273, 277, 323
20,9 formula, 93.252, 269–271, 273
20,14 formula, 93.252, 253, 257, 261–267, 269, 274–275, 277, 286, 318, 323, 324
20,20 formula, 93.252, 286
20S formula, 93.257–258, 266, 277
20T formula, 93.257–258, 266, 269, 277
20U formula, 93.257–258, 266, 277, 323
24,2 formula, 93.252, 253
24,5 formula, 93.252, 253, 266, 269–271, 273, 277, 324
24,9 formula, 93.252, 253, 272
24,14 formula, 93.252, 253, 266, 277
24,20 formula, 93.252, 253, 266, 277, 323–324
Twiddle factor, 93.28
Twinning, microtransform and, 7.49
Twisted nematic cell, 91.246
Two-beam interferometers, 8.34
  signal-to-noise ratio, 8.38
Two-color chromatic aberration, correction of, 14.273–276
Two-defocus method:
  phase determination and, 7.273
  phase problem and, 7.241
  phase problem solution and, 7.225–227
Two-dimensional circulators *See* Circulators
Two-dimensional crystals, negative staining, 8.123
Two-dimensional electron gas, 91.214, 223
Two-dimensional images, reconstruction, 7.293–303
Two-dimensional orientation data, 93.280–283
Two-dimensional phase retrieval, 93.110–112, 131–139, 161–162, 167
Two-dimensional processing, 8.1
Two-dimensional wavelet transforms, 97.197

Two-electrode mirrors, energy analyzers, 89.392, 411–417, 441–458
Two-grid algorithm, 82.330
Two-probe model, 87.149, 155
Two-slab problem, 87.55–61
2,2 formula, 93.233, 252, 253, 261–262, 266, 277
Type, structure, 86.116

U

"Übermikroskop," 10.216, 223
UDUSUPTsup-formulation, 85.23, 26, 34, 36, 63
UHV, 11.48, 103
ULSI *See* Ultralarge-scale integrated chips
Ultimate periodicity, minimax algebra, 90.78–79
Ultracomposition, 89.271, 280, 282
Ultrafine magnetic particles *See* Fine magnetic particles
Ultrahigh-order approximation, canonical aberration theory, 91.1–35
Ultrahigh-order canonical aberration theory, 97.360–406
Ultrahigh-vacuum electron microscope:
  Japanese contributions, 96.716
  Vacuum Generators Company:
    niche marketing, 96.532–533
    scanning electron microscope development, 96.508, 519
Ultralarge-scale integrated chips (ULSI), 89.208–210, 215–217
Ultramicrotomy:
  instrument development, 96.386, 461–462
  Japanese contributions, 96.725, 778–781
  Scandinavian contributions, 96.306, 310
Ultrasonic holograms, use of mirror EM, 4.258
Ultrasonic "ophthalmoscope," 11.167
Ultrathin film, MBE, 89.97
Ultraviolet-excited fluorescence, biological stains, 8.53

Ultraviolet microscopy, 10.9
Umbra transform, 89.366
Uncertainty, 88.248; 94.294–295
  crystal-aperture scanning transmission electron microscopy, 93.57–58, 64
  fuzzy set theory, 89.256
  informational uncertainty, 97.10
  measure, 91.37
  measures, 88.251–260; 91.37
  principal, 88.303, 318
  rough sets, 94.181
Uncertainty principle:
  information approach, 90.165–169, 187–188
  quantum neural computer, 94.304, 305, 307–309
  scale spaces and, 99.3
Uncorrelated covariance, 85.22
Uncorrelated noise, 85.24, 36
Underfiltering, mathematical morphology, 89.351, 353–354
Underlying finite graph, 90.38
Unentangled tree, 84.22
Unicellular organisms, radiation damage, 5.336, 337, 343
UNICOS (software), 85.273
Unified r,s-divergence measures
  Fisher measure of information, 91.115–120
  probability of error, 91.120–132
Unified r,s-entropy, 91.41–42
  bivariate, 91.95–96, 98–102
  multivariate, 91.95, 102–107
  properties, 91.73–75
Unified r,s-inaccuracies, 91.41, 42
  optimization, 91.71–73
Unified r,s-information measures, 91.41–75
  applications, 91.110–132
  Fisher measure of information, 91.115–120
  Markov chains, 91.111–112
  probability of error, 91.120–132
  composition relations, 91.48–53
  convexities, 91.46–48, 62–70

  majorization, 91.47
  in pairs, 91.68–70
  pseudoconvexity, 91.47, 66–67
  quasiconvexity, 91.47, 67
  Schur-convexity, 91.48, 67–68, 70
  inequalities among, 91.57–62
  M-dimensional, 91.38–40, 75–95
  mutual information, 91.107–110
    Markov chains, 91.111–112
  Shannon–Gibbs inequalities, 91.42, 53–57
  unified r,s-entropy, 91.41–42
    bivariate, 91.95–96, 98–102
    multivariate, 91.95, 102–107
    properties, 91.73–75
  unified r,s-inaccuracies, 91.41, 42
    optimization, 91.71–73
Unified r,s-information radii, 91.113
  M-dimensional, 91.76–77
Unified r,s-measures, 91.57–62
  M-dimensional generalization, 91.8295
Unified r,s-mutual information, 91.107–110
  Markov chains, 91.111–112
Unified r,s-relative information, 91.41, 43
  Kullback–Leibler, 91.60
Uniformity, orientation analysis, 93.282
Uniformity enhancement, 92.9–19
  with additive noise or texture, 92.10–13
  with multiplicative noise or texture, 92.13–14
  nonlinear mean filter, 92.18
  rank-order filtering, 92.14–18
    adaptive quartile filter, 92.14–16
    center-weight median filter, 92.17
    composite enhancement filter, 92.16–17
    iterative noise peak elimination filter, 92.17–18
  streak suppression, 92.19
Unigrid, 82.346
Union, τ-mapping, 89.338–339, 342–343
Union Optical Company "Farom" microscope, 5.33–35

Uniqueness:
  of eddy current field, 82.22
  of electrostatic field, 82.6
  of magnetostatic field, 82.8
  of phase solution, 7.238–246
  of scalar potential:
    electric, 82.11, 27
    magnetic, 82.13, 15, 29
  of static current field, 82.9
  of vector potential:
    current, 82.20
    magnetic, 82.17, 26, 27
Unitary matrix, 86.212
United Kingdom:
  electron microscopy, 10.221–224, 254–256
  electron physics during World War II, 91.260–274
  integrated circuits, 91.150–151, 168–169
  radioastronomy, 91.285–290
  semiconductor industry, 91.149–151, 168–169, 177–178, 181
  telecommunications, 91.191–194, 196–205
  television, 91.193
United States:
  Electron Microscopy Society of America See Electron Microscope Society of America
  history of electron microscopy
    commercial microscopes, 96.354–356, 360–362, 365–366
    computer technology, 96.368–369
    electron optics, 96.347–348
    high-resolution electron microscopy, 96.367, 369
    high-voltage electron microscopy, 96.366, 369
    information sources, 96.348–352
    materials science, 96.363–364
    meetings, 96.348, 357
    microscope construction, 96.353–354
    research sites, 96.352–353, 356, 367
    significant events, 96.375–381
    specimen preparation, 96.362–363

radioastronomy, 91.287, 288
semiconductor industry, 91.146, 149, 168–169
Universal conductance fluctuations, mesoscopic devices, 91.216–218
Universal field, 94.307
Universality, of description tool, 94.326
Universal Line-standards Comparator, 8.27
Universal stage, for orientating crystals, 1.46
University of Tokyo, microscope development, 96.251–256
Unsharp masking, weighted, 92.24–25
Upper approximation, of a set, 94.157–165
Upper boundary, 94.164
Upper substitutional decision, 94.185
Upper-triangular matrix, 85.22–23, 27–28, 30, 34–35, 37, 43, 45, 47, 51–52; 90.53–54
Uranium oxide, irradiation, 7.293
Uranyl acetate staining, 2.276, 280, 326; 8.91–93, 94–95

V

Vacancies, 11.80; 87.219, 221
Vacuum:
  mirror electron microscopy, 4.215
  scanning electron microscopy, 83.211, 228, 256–257, 259
  at low beam voltage, 83.221
  breakdown, 83.27, 49, 80
  contamination, 83.213, 237, 239
  differential pumping, 83.230
  electron gun, 83.229
  low temperature, effect of, 83.213, 237, 240
  modifications for LVSEM, 83.230
  molecular drag pumps, 83.230, 256, 259
  oil-free, 83.229–230
  van der Waals force, 83.213
Vacuum evaporation, support film preparation by, 8.120

Vacuum field emission triode (VFET), 83.72
Vacuum fluorescent display, 91.236
Vacuum Generators Company:
    niche marketing, 96.532–533
    scanning electron microscope development, 96.508, 519
Vacuum microelectronics, 83.90–91
    devices:
        atmospheric operations, 83.85–87
        displays, 83.79–82
        electron guns, 83.75–77
        integrated circuits, 83.67–75
        ion sources, 83.82–85
        microwave power tubes, 83.77–78
        scanning tunneling microscope, 83.87–90
    electron sources, physics of, 83.6–35
    history, 83.2–5
    microfield emission sources, 83.36–67
VAIL *See* Variable axis immersion lens
VAL *See* Variable axis lens
Valence band, 83.2, 10, 21
Valence-band excitation, 86.187
Valine crystals, 5.309, 314, 316
Value functions, 94.3, 45
Value set, 90.358
Van Cittert–Zernike theorem, 7.112, 114
Vander Lugt filters, 8.22
van der Waals forces, 87.53–102; *See also* Hamaker constant
    absorbed surface layers, 87.90–92
    application, molecular-scale analysis and surface manipulation, 87.97–100
    bead–substrate interaction, 87.99–100
    continuum, 87.117
    description, 87.53–55
    dielectric contributions, 87.72–87
    differential power law index, 87.70
    dispersion interaction, 87.68–69
    effective measure of curvature, 87.66
    excess dielectric polarizability, 87.62
    four-slab arrangement, 87.90–91
    intermolecular force, 87.61
    interpolation between asymptotic regimes, 87.60
    lateral resolution, 87.71
    macroscopic consequence, 87.54
    metal probe interaction with metal and mica substrate, 87.88–90
    "molecular tip array," 87.99–100
    multipole contributions, 87.96
    noncontact microscopy, 87.101
    observability, 87.87–89
    particle–substrate dispersion interaction, 87.68
    pressure, 87.56, 58–60
        four-slab, 87.91
        as function of separation, 87.86–87
        interaction with ionic pressure, 87.105–106
        two-slab, 87.87, 89
    probe geometry effect, 87.65–72
    retardation effect onset, 87.89
    retardation wavelength, 87.83–86
    retarded vacuum dispersion force, 87.95–96
    size, shape, and surface effects, 87.92–97
    sliding process, 87.97
    sphere-slab arrangement, 87.93–94, 96
    theory limitations, 87.92–97
    tip–particle interaction, 87.97, 99
    transition distance, 87.83
    transition to renormalized molecular interactions, 87.61–65
    two-slab problem, 87.55–61
    two-sphere configuration, 87.94, 96
Vapor phase, 83.14
Variability of prototypes, 84.238, 245
Variable axis immersion lens (VAIL), 12.125, 135, 190, 192, 200; 13.160; 83.161, 170–172
Variable axis lens (VAL), 12.125; 83.160, 161
Variable color-amplitude phase system, 6.103, 104
Variable-phase complex filters, 8.22
Variable-phase contrast systems, 8.18
Variable range hopping, 82.206

Variable shadow procedure, electron gun geometrical properties determination by, 8.147–148
Variable space memory, 87.290
Variational analysis:
    complex media, 92.142–145
    gyroelectric-gyromagnetic, 92.136–138
    isotropic gyroelectric and gyromagnetic, 92.139–141
    propagation constant, 92.93
Variational correction, 82.212
Variational method, 82.212
Variational principle, 82.239
Variation principles, 86.248
Varicella virus, three-dimensional reconstruction, 7.368
Varicolor phase contrast device, 6.95
Vascular bundle cells, 5.140, 146
VDU *See* Visual display unit
Vector adaptive predictive coding, 82.173
Vector computers, 85.25
Vector convolution, 94.11
Vector excitation coding, 82.173
Vector field, 84.136, 84.138, 84.170, 84.184 ff.
    holonomy, 84.138
    prolongations, 84.138, 141, 187
Vector filter, 94.385
Vector Helmholtz equation:
    chiral media, 92.122–123, 126, 130
    sourceless, 92.135, 159
Vector predictive coding, 82.157
Vector processor, 85.24–26, 32, 62–63
Vector quantization, 82.137; 84.3; 97.51, 193
    adaptive, 82.143
    optimality conditions, 82.137
    suboptimal, 82.142
    successive approximation wavelet vector quantization (SA-W-LVQ), 97.191–194
        coding algorithm, 97.220–226
        image coding, 97.226–232
        theory, 97.193–220
        video coding, 97.232–252
        wavelets, 97.201–202, 203
Vector quantizers (VQs), 84.4–6
    equivalent quantizers, 84.13–14
    residual vector quantizers, 84.1–52
    single-stage quantizers, 84.9–10
Vector residual quantizers, 84.26–29
Vectors:
    eigenvectors, 90.63–67
    minimax algebra, 90.5
Vector space:
    finite abelian group, 93.7–8
    group algebra elements, 94.28–29
Vector state function, 94.385
Vector sum excitation linear prediction (VSELP), 82.176
Vector transform quantization, 82.159
Vector wavelet transform, 97.202
Vedic cognitive science, 94.264–265, 267
Velocity:
    of blood, 1.7, 8, 10, 16, 23, 24, 25, 26, 29
    χ function, 94.328
    of moving particles, 1.1, 28
Vertical recording media, 87.164–167
Vertical semiconductor quantum devices, resonant tunneling devices, 89.121–142
Vertical stray field, 87.142–144, 152
Very high-speed integrated circuits (VHSIC), 89.208–209
Very large instruction word (VLIW) architecture, 87.269, 270
Very short-term memory, 94.289
Vestopal, 2.320
VFET *See* Vacuum field emission triode
VG scanning electron microscopy, 93.79, 80, 82, 106
VHSIC *See* Very high-speed integrated circuits
Vibrations, scanning transmission electron microscopy, 83.257; 93.86
Vickers Image-Shearing Module Mark I, 9.232–234
    antipolarizing system, 9.234
    brightness equalizations, 9.233–234
    color filter, 9.234

drive mechanism, 9.234
shutter, 9.234
zero shear correction, 9.233
Vickers Image-Shearing Module Mark II, 9.234–236
digital readout unit, 9.236
flexure pivot, 9.235
lens system, 9.236
optical system, 9.235–236
Vickers integrating microdensitometer, 10.9
accuracy, 6.160, 161
calibration, 6.159
curvilinear scan, 6.158, 166
development, 6.135, 152, 164–168
electronic system, 6.154
fast prism, 6.140
glare, 6.155, 156, 161
illumination, 6.142
integrated circuit, 6.159
linearity, 6.161, 162
masking system, 6.147–149
microscope stage control, 6.164
monochromator, 6.150–152
optical density, 6.147
optical design, 6.136, 137
optical system, 6.142–145
prototype, 6.155–163
reproducibility, 6.160, 161
resolution, 6.144, 160, 167
scan area, 6.164, 168
scanning:
  deflected mirror scanner, 6.164–168
  electromagnetic scanner, 6.164–166
  nodding mirror scanner, 6.165
  rectilinear scan, 6.158, 166
scanning spot size, 6.156, 157
scanning system, 6.136, 137–142
scan speed, 6.168
sensitivity range, 6.159
slow prism, 6.141, 142, 157
spectral response characteristics, 6.162, 163
spurious density, 6.148
star collimator, 6.142, 143, 144

transmission–absorbance conversion system, 6.145–147
uniformity of field, 6.145
Vickers microscopes, 5.36
Video coding, low-bit-rate, 97.232–252
Video image processor, 87.286–287
Video signals, 84.328; 97.192
Vidicon tubes, 6.24; 9.40, 41–42, 44, 280–281, 343
Viewing parameter, 85.211
Viewing transformation, 85.117
Vignetting, 10.71; 12.52
Virtual diffuse scattering, 90.348–349
Virtual immersion lens, 12.189
Virtual process, 86.183
Virtual reality, 85.82; 91.208–209, 244, 255–256
Virus *See* Viruses; Virus particle
Virus crystal, 6.229, 239, 258
Viruses, 5.305; 6.228; 10.312
  bacterial, 6.228, 241, 242
  biochemical analysis, 6.239
  buffer sensitivity, 6.236, 237
  capsid, 6.240, 241, 261, 268
  capsomere, 6.240, 267, 268
  collapse, 6.237
  core, 6.240
  cryoelectron microscopy, 96.739
  electron microscopy, 6.227–230, 229, 230, 232, 233, 236, 237, 239, 239–243, 261; 7.288
    analogs for interpretation, 6.258–261
    buffer and, 6.236, 237
    duplication from models, 6.260
    hidden periodicity, 6.243
    image reconstruction, 6.242
    integration, 6.243, 245, 246, 247
    interference, 6.242
    moire pattern, 6.242
    noise, 6.242
    photographic averaging, 6.243–250
    resolution, 6.242
    signal-to-noise ratio, 6.242, 243
    stain and, 6.233, 234
    superimposition, 6.242, 243

Viruses, *continued*
  virus symmetry pattern, 6.240, 242
  envelope, 6.240
  filament, 6.240, 241
  fixed, 6.236, 237
  freeze-etch replication, 8.93
  helical, 6.240, 260
  hydrodynamic study, 6.239
  icosahedral, 6.240, 260, 265
  immunoelectron microscopy, 96.740
  negative staining, 6.227, 228, 229–238; 7.288
    carbon preparation, 6.268–272
    material, 6.261–264
    with potassium phosphotungstate, 8.91, 94–95
    stain preparation, 6.230–233, 236–238, 242, 255, 257, 258
  nomenclature, 6.239–242
  nucleic acid, 6.239
  nucleocapsid, 6.240, 260, 268
  optical diffraction patterns, 6.251, 253, 254, 256, 258, 265, 266
  osmotic sensitivity, 6.236
  pathogenic, 6.233
  protein tube, 6.256, 257
  radiation damage, 5.318
  rapid-freeze method, 96.740–741
  resolution, 6.234
  rod, 6.240, 241
  shell, 6.240
  SiO replication, 8.90–91
  soft, 6.237, 238
  spherical, 6.227; 7.367–373
  structure, 6.239, 240, 241, 242, 250; 7.282, 295
  symmetry pattern, 6.240, 241, 242
  three-dimensional structure, 7.282, 295
  ultrastructural studies:
    France, 96.94–95, 99–100
    Japan, 96.682–683, 735–741
  unfixed, 6.237
  uranyl acetate staining, 8.91–93, 94–95
  X-ray analysis, 6.239, 254, 258, 260

Virus particles:
  disruption, 6.268
  helical model, 6.260
  isolation, 6.229
  linear repeating features, 6.243
  low molecular weight components, 6.268
  model construction, 6.260
  one-sided image, 6.242, 260
  periodicity, 6.243
  phosphorus content, 3.123–126, 152
  in regular crystalline array, 6.261, 265, 267
  rotational symmetry, 6.243
  surface interference patterns, 6.258, 268
  three-dimensional models, 6.258, 260
  two-sided image, 6.242, 258, 260
  X-ray analogue technique, 6.260, 261
Viscosimetry, 2.288, 299, 305
Vision, quantum and neural computing, 94.273, 283–284, 302
Vision modeling:
  Gabor functions, 97.17, 34–37, 41–45
  joint representations, 97.16–19, 37–50
  receptive field, 97.40–44
  sampling, 97.3–4, 45–50
VisTA system, 87.288–291
Visual cortex:
  image representation, 97.37, 39–45
  sampling, 97.45–50
Visual display unit (VDU), 91.238–242, 243
Visual micrometer-microscopes, 8.26
Visual perception, cancellation, 84.138, 141
Visual psychophysics, 97.35, 39, 40, 43
Visual streak image technique, 1.11, 32, 36
VLIW architecture *See* Very large instruction word (VLIW) architecture
Vocoders, 82.148–150
Voltage breakdown, 83.11
Voltage coding, 12.143
Voltage conditioning, 83.6
Voltage contrast, 12.139, 142

Voltage contrast detectors, 12.184, 185
Voltage fluctuations, transfer theory and, 7.132
Voltage measurements, 12.167
Volterra's type, 86.194
Volume calculation, 85.205
Volume data set, 85.89, 206
Volume models, 85.96
von Ardenne, Manfred
   impact of World War II, 96.649–650
   relationship with Ruska, 96.639–641
   scanning electron microscope development, 96.638–639, 643–644
   scanning transmission electron microscope development, 96.644–645
   translation of works into Japanese, 96.240, 649
   transmission electron microscopy contributions, 96.639–641, 645–647
von Borries, Bodo, electron microscopy contributions, 96.794–796
Von Neumann real algebra, 84.272, 284
Voronai diagram, 86.147
Vortex nucleation process, probe-induced, 87.186–187
Voxels, 84.203; 85.115
   address, 85.206
   carryover, 85.182
   filling, 85.182
   model, 85.116, 195, 206
   projection, 85.114, 142
   traversal, 85.127
   visible, 85.138
VQs *See* Vector quantizers
VSELP *See* Vector sum excitation linear prediction

W

Walsh convolution, 94.19
Walsh matrix, 88.15
   binary, 88.20
   conversion between orderings, 88.22
   dyadic-ordered, 88.16, 22
   natural-ordered, 88.17, 21
   sequency-ordered, 88.16, 21
Walsh transform, 88.15
   fast computation algorithm, 88.25
   sinusoidal transforms, 88.15–24
Warp cell, 87.278
Water:
   carbon dioxide bubbles in, 98.44–61
   collision of microdrops, 98.13–21, 22
   fluid models, 98.45–52
   liquid drop formation on a solid surface, 98.30, 32–44
   vapor dropwise condensation studies, 8.80–81, 97–99
Water drop SiO replication, 8.77–79, 90–91
Watershed transform, 99.46, 47–48
Wave aberration, 12.30; 89.8, 9, 21, 48
Wave equations, 90.124
   information approach, 90.161, 164, 184
   multiparticle, 95.364–366
Wavefront division interferometry, 99.194–200
Wavefront reconstruction, direct, 10.148–153
   reversed, 10.153–157
Wave functions:
   elastic, 90.217
   electron, 90.216
   neural processes, 94.263, 298, 301
   rotationally symmetric fields, 13.247–255
   electron charge, conservation of, 13.254–255
   Fraunhofer diffraction, 13.257–261
   Fresnel diffraction, 13.257–261
   ideal imaging, proof of, 13.256–257
   in paraxial condition, 13.255–261
   Schrödinger equation, 13.248–254
Waveguide, 82.4, 38, 39, 40, 45, 61, 85, 86, 87
Wavelength, electron, 2.170; 83.205, 221

Wavelet coefficients:
  scalar quantization, 97.200–201, 202–203
  vector quantization, 97.201–202, 203
Wavelets:
  edge detection, 97.63–64
  signal and image processing, 97.3–5, 11–13, 52–53
Wavelet transforms, 97.12, 52–53, 194
  defined, 97.195
  image compression, 97.52–53, 194, 198–205
  signal description, 94.321
  theory, 97.195–197
  two-dimensional, 97.197
Wave mechanics, 84.262–263, 276
Wavenumber, free space, 82.61, 62
Wave optics:
  differential phase contrast mode, 98.357–358
  Fresnel mode, 98.341–350
  multislice approach, 93.173–216
  relation to wave mechanics, 5.268, 269, 270
Wave packets, coherence and, 7.172
Wave–particle duality, 89.99, 110, 118; 94.295, 296
Wayland–Frasher intravital microscope applications, 6.3, 36
Weakened law of contradiction, 89.260
Weakened law of excluded middle, 89.260
Weak formulation:
  of eigenvalue problem, 82.45
  of second-order elliptic differential equation, 82.43
  of transient problem, 82.44
Weakly definable partition, 94.168
Weakly scattering specimens, transfer functions, 7.122–140
"Weak overlap approximation," 87.108
Weak phase approximation, electron diffraction data, 7.199–211
Weak phase object, phase information, 7.265–269
Weak phase–weak amplitude approximation, 7.211–218
Weak phase–weak amplitude object, phase problems and, 7.269
Weak realization problem, 90.32–33
Weak transitive closure, 90.41–45
Wedges, rims and edge emitters, 83.38–43, 46, 75, 76, 78
Wedge-type interference filters, 8.14
"Wehnelt" cylinder, 8.138; 10.245–247
Wehnelt voltage, electron trajectories and, 8.140–144
Weighted decimation, 93.16
Weighted median filter, 92.17
Weighting filter, 82.154, 155, 163
Weighting function, 82.46, 47, 48, 51, 53, 60, 63
Weingarten operator, 84.191
Welch Direct Torr mechanical pump, 8.75
Wet-cell microscopy, Japanese contributions, 96.786–787
Wet replication techniques, 8.52–53, 72–103
  apparatus, 8.72–77
  diffusion controlled evaporation, 8.80
  evaporant gas selection, 8.77
  future applications, 8.99
  grid preparation, 8.81
  grid surface temperature studies, 8.80
  high resolution studies, 8.80–84
  hydration chamber, 8.72–74
  hydrophilic substrate specimen mounting, 8.81
  microsurface spreading and, 8.93–97
  results analysis and interpretation, 8.97–99
  shape replication by, 8.85–93
  supercooling, test for, 8.79–80
  water thinning, 8.80–84, 97–99
  wet surface freezing, test for, 8.79–80
WF Company, commercial microscope production, 96.429, 452, 458, 476–477, 482–483
Whiskers, 83.16, 54, 55
White blood cell differential count, 5.46–50, 72, 73

Whitening filter, 82.122
White noise, 88.310
White uncertainty processes, 95.196–197
Wide track scanner, 4.362
Wiener filter, 10.12; 93.303–305
Wiener weight vector, 84.266
Wien filter, 11.114; 12.195; 13.107; 86.279
  analysis of, 13.108–113
  geometry of filter and axial fields, 13.109–110
  principle of, 13.108–109
  ray tracing through, 13.110–113
Wigner distribution function, 97.2–3, 9
Wigner–Ville description, signal description, 94.320, 321
Window, mathematical morphology, 89.335
Windowed Fourier transform *See* Short-time Fourier transform
Window tranformation, 89.358
Winograd algorithm, 94.18
Wireframe, 85.102
Wire mesh, dust contaminated Fourier transforms, 8.16–17
WISARD system, 87.287–288
Work function, 12.163, 176; 83.2, 9, 10, 15, 17, 57, 63, 75
  cooling, 83.27, 30
  field electron emission, dipole layer effect on, 8.220
WORMOS (software), 87.282–283
WRITE mechanism, single-electron cells, 89.235–237

X

XSUP#sup invariant reduced transform algorithm, 93.41–42
Xenon, melting point, 98.10
XPS *See* Electron spectroscopy
X-ray absorption near edge structures (XANES), 9.113, 119
X-ray analog technique, for negative stained particles, 6.258, 260, 261

X-ray computed tomography, 14.216
X-ray crystallography:
  phase problem in, 7.189, 245
  phase retrieval, 93.167–168
X-ray detectors, analytical microscopy, 10.258, 259
X-ray dichroism, magnetic microstructure, 98.330
X-ray diffraction, 3.155, 214; 4.113, 115, 118; 13.115, 118
X-ray intensity measurement, counting, 6.283–285
X-ray kymography, 7.74, 75
X-ray mapping, 93.312–319
X-ray microanalysis:
  development in China, 96.834–837
  development in Japan, 96.696, 704
X-ray microprobe analyzer, 10.225
X-ray microscopy, historical note, 10.101–102
X-ray "mirror," LVSEM image, 83.252
X-ray photoelectron microscopy, 10.319–323
X-ray projection microscope, invention, 96.143
X-ray protection, 2.225
X-ray scan color composite, 6.293–295
X-ray signal control, 6.294, 295
X-ray topography, magnetic microstructure, 98.329
X-ray transmission tomography, image reconstruction, 97.159

Y

YBaSUB2subCuSUB3subOSUB7–$x$sub, 14.28–29
  critical field and current density in, 14.31–36
  experiments with, 14.54–55, 57–58
  substitution in, 14.30–31
  in thin film stack pattern, 14.109–110
Yeast, 6.65
  Dutch electron microscopy research, 96.271–273, 276, 281, 283–284

202 INDEX

Yeast, *continued*
 Interphako microscopy, 6.109
 Japanese contribution to ultrastructure, 96.744–745
 phase contrast/fluorescence microscopy, 6.118
 positive/negative phase contrast microscopy, 6.94
 protoplast, LVSEM, 83.244
Yoshioka's coupled equations, 86.184, 195, 203
Young's interference fringes, 10.115–116
Yukawa potential, 82.205, 225

Z

Zak transform, Gabor expansion, 97.29–30
ZAP *See* Zone axis pattern
ZAT *See* Zone axis tunnels
Z-buffer, 85.117, 174
Z contrast *See* Density
Z-contrast imaging, 90.289–293
Zeiss:
 collaboration with AEG, 96.428, 433
 computer control of microscopes, 96.516
 EF4, 96.472–473
 EF5, 96.473
 EF6, 96.473–474, 488
 electron microscope development, 96.448, 458
 EM8, 12.9
 EM9, 96.471–472, 486
 EM10, 96.494–495
 EM109, 96.499
 EM902, 96.515–516
 EM912, 96.516
 energy filter incorporation, 96.792
 Neophot microscope, 5.27–30
 Optovar in tracking microscope, 7.3
 photomicroscope, 1.101
Zener diode, 87.244

Zernike phase contrast method, 5.249; 6.52–62, 67, 89, 90, 95; 10.25; 12.28, 40, 74, 76
Zero field boundaries, in FDM methods, 13.34–35
Zero-field spin splitting, 89.201
Zero information, 90.143–144, 164, 185
Zero input response, 82.164
Zero location method, phase retrieval, 93.112, 118, 167
Zero-order Laue zone (ZOLZ), 90.240
Zero property, Lagrangians, 90.126–127
Zero sheets method, phase retrieval, 93.112, 131, 167
Zero state response, 82.164
Zero-tree root, 97.224
Zero-trees, 97.202–203
Zeta surface potential, 8.112
Ziferan, 8.125
Zinc silicate binary mixture, 10.300–303
ZOLZ *See* Zero-order Laue zone
Zonal filtering, 92.25
Zone axis pattern (ZAP), crystal-aperture scanning transmission electron microscopy, 93.58, 66–73
Zone axis tunnels (ZAT), 93.62–65, 73–79, 85, 107
Zone plates, 1.137, 271
Zoom condenser, 3.2; 83.128
Zoom systems, 3.1
 catalog, 3.13
 design, 3.5
 eyepieces, 3.3, 15
 history, 3.1
 mechanically compensated, 3.8
 objectives, 3.3, 15
 optically compensated, 3.10
Z-transformation, 85.16
Zuniga and Haralick formula, 93.253–267, 269, 277, 323
Zwischen electrode, 11.106

ISBN 0-12-014742-4